Uncovering
Student Ideas in Earth and Environmental Science

32 New Formative Assessment Probes

To Anne—
thanks for all
you do to
support
learning!

Uncovering
Student Ideas in Earth and Environmental Science
32 New Formative Assessment Probes

By **Page Keeley**
and
Laura Tucker

National Science Teachers Association
Arlington, Virginia

National Science Teachers Association

Claire Reinburg, Director
Wendy Rubin, Managing Editor
Rachel Ledbetter, Associate Editor
Amanda O'Brien, Associate Editor
Donna Yudkin, Book Acquisitions Coordinator

ART AND DESIGN
Will Thomas Jr., Director
Cover, Inside Design, and Illustrations by Linda
Olliver

PRINTING AND PRODUCTION
Catherine Lorrain, Director

NATIONAL SCIENCE TEACHERS ASSOCIATION
David L. Evans, Executive Director
David Beacom, Publisher

1840 Wilson Blvd., Arlington, VA 22201
www.nsta.org/store
For customer service inquiries, please call 800-277-5300.

NSTA is committed to publishing material that promotes the best in inquiry-based science education. However, conditions of actual use may vary, and the safety procedures and practices described in this book are intended to serve only as a guide. Additional precautionary measures may be required. NSTA and the authors do not warrant or represent that the procedures and practices in this book meet any safety code or standard of federal, state, or local regulations. NSTA and the authors disclaim any liability for personal injury or damage to property arising out of or relating to the use of this book, including any of the recommendations, instructions, or materials contained therein.

Cataloging-in-Publication Data for this book and the e-book are available from the Library of Congress.
ISBN: 978-1-938946-47-9
e-ISBN: 978-1-68140-007-5

SUSTAINABLE FORESTRY INITIATIVE   Certified Sourcing
www.sfiprogram.org
SFI-00756

Contents

Foreword ix

Preface xi

Acknowledgments xxi

About the Authors xxiii

Introduction 1

Section 1. Land and Water

Concept Matrix 12

Related *Next Generation Science Standards* Performance Expectations .. 13

Related NSTA Resources 13

1   What's Beneath Us? 15

2   What Do You Know About Soil? 19

3   Land or Water? 25

4   Where Is Most of the Fresh Water? 29

5   Groundwater 33

6   How Many Oceans and Seas? 37

7   Why Is the Ocean Salty? 41

Section 2. Water Cycle, Weather, and Climate

Concept Matrix 46

Related *Next Generation Science Standards* Performance Expectations ..47

Related NSTA Resources 47

8   Water Cycle Diagram 49

9   Where Did the Water in the Puddle Go? 53

10   Weather Predictors 57

11   In Which Direction Will the Water Swirl? 61

12   Does the Ocean Influence Our Weather or Climate? 65

13   Coldest Winter Ever! 69

14   Are They Talking About Climate or Weather? 73

15   What Are the Signs of Global Warming? 77

Section 3. Earth History, Weathering and Erosion, and Plate Tectonics

Concept Matrix 84

Related *Next Generation Science Standards* Performance Expectations ..85

Related NSTA Resources 85

16   How Old Is Earth? 87

17   Is It a Fossil? 91

18   Sedimentary Rock Layers 95

19   Is It Erosion? 99

20   Can a Plant Break Rocks? 103

21   Grand Canyon 107

22   Mountains and Beaches 111

23   How Do Rivers Form? 117

24   What Is the Inside of Earth Like? 121

25   Describing Earth's Plates 125

26   Where Do You Find Earth's Plates? 131

27   What Do You Know About Volcanoes and Earthquakes? 135

Section 4. Natural Resources, Pollution, and Human Impact

Concept Matrix 142

Related *Next Generation Science Standards* Performance Expectations 143

Related NSTA Resources 143

28   **Renewable or Nonrenewable?** 145

29   **Acid Rain** 151

30   **What Is a Watershed?** 155

31   **Is Natural Better?** 161

32   **The Greenhouse Effect** 165

Index 171

Foreword

In our efforts to use science to improve the quality of our lives, we have learned that our mental models of physical phenomena can be so deeply anchored as to effectively block learning. From the youngest age, we try to explain natural phenomena—the Sun "comes up" in the morning, the Earth casts a shadow on the Moon, rocks are stronger than plants—and we ask questions. Where do birds go at night? How does an acorn become a tree? Why is the ocean cold in sunny Los Angeles but warm in chilly Maryland? Once we have the answers, we tend to hold them fast, regardless of whether what we "learned" was correct.

Many will remember Science Media Group's 1987 video *A Private Universe*. The video revealed that some of our best-educated college students could not apply the science they had learned about familiar occurrences such as the changing seasons or phases of the Moon. The issue identified in the video and in very substantial research is that the presence of misinformation can prevent correct information from taking root. When teaching science, simply presenting the evidence for a more scientifically accurate explanation is not enough. Misconceptions must be explicitly identified to facilitate learning.

The *Uncovering Student Ideas in Science* series addresses this critically important step in science education. With her engaging volumes, Page Keeley gives teachers the tools they need to identify their students' and their own misunderstandings at the beginning of instruction. To ensure deeper learning, she follows up the accessible probes with related research and suggestions for instruction and assessment. *Uncovering Student Ideas in Earth and Environmental Science: 32 New Formative Assessment Probes* is the 10th book of the series

and focuses on areas of science where we all have our own misconceptions, such as the formation of rock and soil, the processes of weather and climate, the water cycle, the saltiness of the ocean, and the history of Earth.

As you dive into these probes with your students, I encourage you to keep the importance of addressing misconceptions in mind. The Age of Enlightenment taught us that the use of evidence, logic, and reason—cornerstones of scientific investigation—influences how we understand our world. So ensuring that students have a good understanding of science is far more important than focusing on just memorizing "facts." And given the global challenges we face, science education is fundamental to our survival as a society—and maybe even as a species.

Today, we better understand that human activities impact the whole planet (an idea that many of us find hard to accept). Our role as stewards of the environment is thus more critical than the traditional environmentalist message; the habitat, if not the viability, of our species could be at stake. *Uncovering Student Ideas in Earth and Environmental Science* will help you guide your students in making science-based decisions as responsible members of society. I hope that your emphasis on right scientific thinking will lead to right environmental doing.

—David L. Evans, PhD
Executive Director
National Science Teachers Association

Reference

Harvard-Smithsonian Center for Astrophysics, Science Education Department, Science Media Group. 1987. *A private universe*. Film available at *www.learner.org/resources/series28.html*.

Preface

This book is the 10th book in the *Uncovering Student Ideas in Science* series and the first one specifically targeting Earth and environmental science. Like its predecessors, this book provides a collection of unique questions, called formative assessment probes, designed to uncover preconceptions students bring to their learning, as well as identify misunderstandings students develop during instruction that may go unnoticed by the teacher. Each probe is carefully researched to surface commonly held ideas students have about the phenomenon or scientific concept targeted by the probe. Each probe includes one scientifically best answer, along with distracters designed to reveal common, research-identified alternative conceptions held by children and adults.

The 32 probes in this book uncover students' thinking about several of the big ideas in Earth and environmental science. Many of the probes are designed to uncover pre-existing ideas, often developed before the concept or idea is even taught. Therefore, we avoid the use of technical terminology in a probe and instead use everyday language students are familiar with in order to uncover their conceptual ideas that do not depend on knowing vocabulary. Some of the probes are intended for use after a concept or idea has been introduced, such as the collection of plate tectonic probes. For example, students may first need to learn the idea that Earth is composed of several plates before probing students' ideas about the characteristics of Earth's plates.

It is impossible to cover all Earth and environmental science ideas in one book. For this first volume in Earth and environmental science, we chose to focus primarily on ideas associated with strongly held misconceptions that follow students from one grade level to the next, often into adulthood. You may wonder why some ideas such as the rock cycle, energy in the Earth system, flow of matter and energy through ecosystems, and atmospheric ideas are not included in this book. Some of these ideas will be included in other books in the *Uncovering Student Ideas in Science* series. Energy in the Earth system will be covered as a crosscutting concept in the future book, *Uncovering Student Ideas about Matter and Energy.* For the environmental science probes, we chose to focus primarily on natural resources and human impact. Probes related to matter and energy in ecosystems and ecosystem dynamics are included in the collection of life science probes in the other books in this series. *Uncovering Student Ideas in Life Science, Volume 2,* to be released in 2017, will contain additional ecosystem-related probes.

Other *Uncovering Student Ideas in Science* Books That Include Earth and Environmental Science–Related Probes

The following is a description of the other books in the *Uncovering Student Ideas in Science* series to date (2016) that include probes related to Earth and environmental science.

Uncovering Student Ideas in Science, Volume 1 (Keeley, Eberle, and Farrin 2005): This first book in the series contains 25 formative assessment probes in life, physical, Earth, and space science. The introductory chapter of the book provides an overview of what formative assessment is and how it is used. Earth and environmental science probes in this book, along with suggested grade levels and related concepts, include the following:

Preface

- "Wet Jeans" (grades 3–12): water cycle and evaporation
- "Beach Sand" (grades 5–12): weathering, erosion, deposition, and beach formation
- "Mountain Age" (grades 5–12): mountain formation

Uncovering Student Ideas in Science, Volume 2 (Keeley, Eberle, Tugel 2007): This second book in the series contains 25 formative assessment probes in life, physical, Earth, and space science. The introductory chapter of this book describes the link between formative assessment and instruction. Earth and environmental science probes in this book, along with suggested grade levels and related concepts, include the following:

- "Is It a Rock? Version 1" (grades 2–5): rock and rock sizes
- "Habitat Change" (grades 3–8): adaptation and habitat change
- "Is It a Rock? Version 2" (grades 3–8): concept of a rock and natural versus human-made rocks
- "Mountaintop Fossil" (grades 3–8): fossils, mountain formation, uplift, and plate tectonics
- "Giant Sequoia Tree" (grades 6–12): photosynthesis and carbon cycle

Uncovering Student Ideas in Science, Volume 3 (Keeley, Eberle, Dorsey 2008): This third book in the series contains 25 formative assessment probes in life, physical, Earth, and space science. It also contains three nature of science probes on hypotheses, theories, and how scientists do their work. The "Is It a Theory?" probe can be combined with the collection of plate tectonics probes. The introductory chapter of the book describes ways to use the probes and student work for professional learning. Earth and environmental science probes in this book, along with

suggested grade levels and related concepts include the following:

- "Rainfall" (grades 3–8): rain, precipitation, and weather
- "Rotting Apple" (grades 3–8): decay and decomposers
- "What Are Clouds Made of?" (grades 3–8): clouds and water cycle
- "Where Did the Water Come From?" (grades 3–12): water cycle and condensation
- "Earth's Mass" (grades 5–12): flow of matter through ecosystems and conservation of matter

Uncovering Student Ideas in Science, Volume 4 (Keeley and Tugel 2009): This fourth book in the series contains 25 formative assessment probes in life, physical, Earth, and space science. It also includes two probes that target the crosscutting concepts of models and systems. The introductory chapter of this book describes the link between formative and summative assessment. Earth and environmental science probes in this book, along with suggested grade levels and related concepts include the following:

- "Where Does Oil Come From?" (grades 3–8): fossil fuels
- "Where Would It Fall?" (grades 3–8): land-water distribution
- "Camping Trip" (grades 5–12): Earth's warming and cooling and radiant energy
- "Global Warming" (grades 6–12): global warming and human impact

Uncovering Student Ideas in Life Science, Volume 1 (Keeley 2011b): This sixth book in the series, as well as the first one in the series of life science probes, contains 25 life science formative assessment probes. The introductory chapter of this book describes how formative assessment probes are used in a life science context. Environmental science probes in this book, along with suggested grade levels and related concepts include the following:

- "No More Plants" (grades 2–8): role of producers, food chains, and food webs
- "Is It a Consumer?" (grades 3–8): consumer, food web, and food chain
- "Changing Environment" (grades 5–12): adaptation and ecosystem change
- "Food Chain Energy" (grades 5–12): producers, consumers, and flow of energy
- "Ecosystem Cycles" (grades 6–12): matter cycles and energy flows

Uncovering Student Ideas in Astronomy (Keeley and Sneider 2012): This seventh book in the series contains 45 astronomy formative assessment probes. Many Earth science teachers also teach space science and can use these probes to address the space sciences section of their curriculum. The introductory chapter of this book describes how formative assessment probes are used to understand students' mental models in astronomy. In addition to the astronomy probes, probes included in this book that address students ideas about the nature of planet Earth include the following:

- "Is the Earth Really Round?" (grades 2–5): concept of a spherical Earth
- "Where Do People Live?" (grades 2–5): concept of a spherical Earth
- "Falling Through Earth" (grades 6–8): Earth's gravitational attraction

Uncovering Student Ideas in Primary Science, Volume 1 (Keeley 2013): This eighth book in the series contains 25 formative assessment probes designed for K–2 students. The probes are for early or nonreaders as well as English language learners. They can also be used in grades 3–5 to check for prior knowledge. The probes are visual in nature and designed to be

used in a talk format. The introductory chapter focuses on how to use the probes to support science talk and how science talk supports students' thinking. Earth and environmental science probes in this book, along with suggested grade levels and related concepts, include the following:

- "Describing Soil" (grades K–5): soil
- "Is a Brick a Rock?" (grades K–5): rock and natural versus human-made materials
- "What Makes Up a Mountain?" (grades K–5): mountains and rock

Format of This Book

This book contains 32 probes for grades 3–12 and is organized in four sections: Section 1, "Land and Water" (7 probes); Section 2 "Water Cycle, Weather, and Climate" (8 probes); Section 3 "Earth History, Weathering and Erosion, and Plate Tectonics" (12 probes), and Section 4 "Natural Resources, Pollution, and Human Impact" (5 probes). The format is similar to the other nine volumes in the *Uncovering Student Ideas in Science* series. The introductory chapter describes how to use the probes and provides an overview of teaching and learning related to Earth and environmental science. Each section begins with a concept matrix that lists the main concepts that each probe addresses. The matrix also lists the related performance expectations from the *Next Generation Science Standards* (*NGSS*) by grade level and related National Science Teachers Association (NSTA) resources, such as journal articles, books, content webinars, and science objects. These resources provide materials for teachers who wish to extend their learning. The Teacher Notes are one of the most important components of the book and should always be read before using a probe. The following pages describe the features of the Teacher Notes that accompany each probe in this book.

Preface

Purpose

This section describes the purpose of the probe—that is, what you will learn about your students' ideas if you use the probe. It begins by describing the overall concept the probe elicits, followed by the specific idea the probe targets. Before choosing a probe, you must understand what the probe is intended to reveal. Taking time to read the purpose will help you decide if the probe will elicit the information you need to understand your students' thinking.

Type of Probe

This section describes the format of the probe. The probes in this series use 10 different formats. Some of the more common formats are justified lists, friendly talk, and opposing views. The format of a probe is related to how a probe is used. The snapshot vignettes in the Introduction (pp. 1–10) illustrate how a format informs the use of a probe.

Related Concepts

Each probe is designed to target one or more concepts that are often used across multiple grade levels. A concept is a one-, two-, or three-word mental construct used to organize the ideas the probe addresses. Most of these concepts are included in core disciplinary ideas. The concepts are also included on the matrix charts that precede the probes for each section.

Explanation

The best answer choice is provided in this section. We use *best* answer rather than *correct* answer because the probes are not intended to pass judgment on students. Instead, they are used to encourage students to reveal their thinking without the worry of being "wrong." Sometimes there is no single "right" answer because the probe is designed to uncover different ways of thinking. The *best* answer is the one that scientifically addresses the purpose and intent of the probe.

A brief scientific explanation accompanies each probe and clarifies the scientific content that underlies the probe. The explanations are designed to help you identify what the most scientifically acceptable answers are, as well as clarify any misunderstandings about the content. The explanations are not intended to provide detailed background knowledge about the content. They are provided to support teachers' content knowledge; although in some cases, the explanations can be shared with upper middle and high school students as written. Some elementary and middle school science teachers have limited coursework or professional development in science, and some high school instructors teach Earth or environmental science outside of their science major. Therefore, the explanations are carefully written to avoid highly technical language so that you do not have to be a science specialist to understand them. At the same time, the explanations try not to oversimplify the science. Rather, they provide the concise information a science novice would need to understand the content he or she teaches related to the probe. If you need additional background information regarding the content of the probe, refer to the NSTA resources listed for each section to build or enhance your content knowledge.

Administering the Probe

Intended grade levels and suggestions are provided for administering the probe to students, including response methods, ways to use props, the way to demonstrate the probe scenario, modifications for different learners, or use of different formative assessment classroom techniques (FACTs) to gather the assessment data. FACTs are described in the Introduction on pages 3–9.

Preface

Related Core Ideas

This section identifies the learning goals described in the two national documents used to develop the learning goals in most states' standards and curriculum materials—the revised, online version of *Benchmarks for Science Literacy* (AAAS 2009) and *A Framework for K–12 Science Education: Practices, Crosscutting Concepts, and Core Ideas* (NRC 2012), of which the disciplinary core ideas were used to develop the *Next Generation Science Standards* (NGSS Lead States 2013). Because those are the primary source documents on which almost all state standards are or will be based after they are revised, it is important to look at the related learning goals in these documents. Because the probes are not designed as summative assessments, the listed learning goals are not to be considered alignments, but rather ideas that are related in some way to the probe. Additionally, the performance expectations related to probes in each section are listed under the concept matrices at the beginning of each section.

Core ideas across grade spans are included in this section. The ideas are included because seeing the related idea that precedes your grade level is useful when using the probe, as well as seeing the core idea that builds on the probe at the next grade level. In other words, teachers can see how the foundation they are laying relates to a spiraling progression of ideas as students move from one grade level to the next.

Related Research

Each probe is informed by related research when available. Three comprehensive research summaries commonly available to educators are the following: Chapter 15 in the *Benchmarks for Science Literacy* (AAAS 1993), Rosalind Driver's *Making Sense of Secondary Science: Research Into Students' Ideas* (Driver et al. 1994), and recent summaries in the *Atlas of Science Literacy* (AAAS 2007) were drawn on for the research summaries. In addition, recent research from science education journals is cited where available. Although many of the research citations describe studies that have been conducted in past decades and studies that include children in not only the United States but also other countries, most of the results of these studies are considered timeless and universal. Whether students develop their ideas in the United States or other countries, research indicates that many of these commonly held ideas are pervasive regardless of geographic boundaries and societal and cultural influences.

Although your students may have had different backgrounds, experiences, and contexts for learning, the descriptions from the research can help you better understand the intent of the probe and the kinds of thinking your students are likely to reveal when they respond to the probe. The research also helps you understand why the distracters are written a certain way. As you use the probes, we encourage you to seek new and additional published research, engage in your own action research to learn more about students' thinking, and share your results with other teachers to extend and build on the research summaries in the Teacher Notes. To learn more about conducting action research using the probes, read the *Science and Children* article "Formative Assessment Probes: Teachers as Classroom Researchers" (Keeley 2011b), or read Chapter 12 in the book *What Are They Thinking? Promoting Elementary Learning Through Formative Assessment* (Keeley 2014b).

Suggestions for Instruction and Assessment

Uncovering and examining the ideas children bring to their learning is considered diagnostic assessment. Diagnostic assessment becomes formative assessment when the teacher uses the assessment data to make decisions about

Preface

instruction that will move students toward the intended learning target. Thus, for the probe to be considered a formative assessment probe, the teacher needs to think about how to design, choose, or modify a lesson or activity to best address the preconceptions students bring to their learning or the misunderstandings that might surface or develop during the learning process. As you carefully listen to and analyze your students' responses, the most important next step is to choose the instructional path that would work best in your particular context according to the learning goal, your students' ideas, the materials you have available, and the different types of learners you have in your classroom.

The suggestions provided in this section have been gathered from the wisdom of teachers, the knowledge base on effective science teaching, and research on specific strategies used to address commonly held ideas and conceptual difficulties. These suggestions are not lesson plans, but rather brief recommendations that may help you plan or modify your curriculum or instruction to help students move toward learning scientific ideas. It may be as simple as realizing that you need to provide a relevant, familiar context, or there may be a specific strategy, resource, or activity that you could use with your students.

Learning is a very complex process and most likely no single suggestion will help all students learn. But that is what formative assessment encourages—thinking carefully about the instructional strategies, resources, and experiences needed to help students learn scientific ideas. As you become more familiar with the ideas your students have and the multifaceted factors that may have contributed to their misunderstandings, you will identify additional strategies that you can use to teach for conceptual change and understanding. In addition, this section also points out other related probes in the *Uncovering Student Ideas in Science* series that can be modified or used as is to further assess students' conceptual understanding.

When applicable, the Suggestions for Instruction and Assessment section includes safety notes for the proposed activities and investigations. These guidelines need to be adopted and enforced to provide for a safer learning and teaching experience. Teachers should also review and follow local polices and protocols used within their school and school district. For additional safety information, read NSTA's "Safety in the Science Classroom" article (*www.nsta.org/pdfs/SafetyInTheScienceClassroom.pdf*) or visit the NSTA Safety Portal (*www.nsta.org/portals/safety.aspx*).

References

References are provided for the information cited in the Teacher Notes, including the original article referenced in the research summaries.

Formative Assessment Probes in the Elementary Classroom

Formative assessment is an essential feature of a learning-focused elementary science environment. To help teachers learn more about using formative assessment probes with elementary students to inform instruction and promote learning, NSTA's elementary science journal *Science and Children* publishes a monthly column by the author titled, "Formative Assessment Probes: Promoting Learning Through Assessment." Your NSTA membership provides you with access to all of those journal articles, which NSTA has archived electronically. Go to the *Science and Children* website at *www.nsta.org/elementaryschool*. Scroll down to the journal archives, and enter "formative assessment probes" in the keyword search box. This will pull up a list of all of Page Keeley's column articles. You can save the articles in your library in the NSTA Learning Center or downloaded them as a pdf.

Table 1 lists the journal issue, title of the column, and topic of the column for the articles that have been published to date related to Earth and environmental science. Check back regularly as more articles are added. Professional developers and facilitators of professional learning communities can also use the articles to engage instructors in discussions about teaching and learning related to the probes and the content they teach. In addition, several of the articles are provided in chapter form, along with a link to the probe and discussion questions for professional learning groups in *What Are They Thinking?* (Keeley 2014b).

Table 1. Earth and Environmental Science Formative Assessment Probes: Promoting Learning Through Assessment

Issue	Title	Topic
September 2010	"Doing Science"	Scientific method and how misuse of the "scientific method" affects students' ideas related to the nature of science
December 2010	"To Hypothesize or Not"	Hypothesis making and misconceptions teachers have about the nature of science that can be passed on to students
November 2011	"Teachers as Researchers"	Biological conception of an animal and how formative assessment probes can be used to engage in teacher action research
April/May 2012	"Food for Plants: A Bridging Concept"	Understanding food, photosynthesis, needs of plants; using bridging concepts to address gaps in learning goals; and understanding students' common sense ideas
July 2012	"Where Did the Water Go?"	Using the water cycle to show how a probe can be used to link a core content idea, a scientific practice, and a crosscutting concept
December 2012	"Mountain Age: Creating Classroom Formative Assessment Profiles"	Understanding weathering and erosion and organizing student data using a classroom profile for instructional decisions and professional development
March 2013	"Habitat Change: Formative Assessment of a Cautionary Word"	Adaptation and how formative assessment helps teachers be more aware of the language they use when teaching concepts such as adaptation
April 2013	"Is It a Rock? Continuous Formative Assessment"	Concept of a rock, natural versus human-made materials, and the Group Frayer Model for continuous assessment
September 2014	"Is It a Theory? Speaking the Language of Science"	Scientific theories and how colloquial language affects our understanding of what a scientific theory is
March 2015	"Soil and Dirt: The Same or Different?"	Soil and how our use of everyday language affects understanding of science concepts
April 2015	"No More Plants!"	Understanding producers, food chains, and food webs and uncovering students' ideas about interdependency and ecosystem change
October 2015	"Wet Jeans"	Understanding evaporation and the water cycle and using real world phenomena to uncover ideas
December 2015	"Mountain Top Fossil: A Puzzling Phenomenon"	Understanding how Earth's surface changes over time using a puzzling phenomenon

Preface

Formative Assessment Reminder

Now that you have the background on the probes and the Teacher Notes in this new book, let's not forget the formative purpose of these probes. Remember that a probe is not formative unless you use the information from the probe to modify, adapt, or change your instruction so that students have the opportunity to learn the important scientific ideas necessary for achieving scientific literacy. As a companion to this book and all the other volumes, NSTA has co-published the book *Science Formative Assessment: 75 Practical Strategies for Linking Assessment, Instruction, and Learning* (Keeley 2008, 2015) and *Science Formative Assessment: 50 More Practical Strategies for Linking Assessment, Instruction, and Learning* (Keeley 2014a). In these books, you will find a variety of strategies to use, along with the probes to facilitate elicitation, support metacognition, spark inquiry and investigation, encourage discussion, monitor progress toward conceptual change, encourage feedback, and promote self-assessment and reflection. In addition, these strategies provide opportunities for students to use scientific practices such as modeling, designing investigations, argumentation, and explanation construction.

Finally, the ultimate purpose of formative assessment is to break away from teaching and assessing disconnected facts to support conceptual learning of science. Because conceptual change is the underpinning of the *Uncovering Student Ideas in Science* series, we highly recommend the book *Teaching for Conceptual Understanding in Science,* which includes chapters on understanding the nature of students' thinking, instructional strategies that support conceptual change, and content that links assessment, instruction, and learning (Konicek-Moran and Keeley 2015).

References

American Association for the Advancement of Science (AAAS). 1993. *Benchmarks for science literacy.* New York: Oxford University Press.

American Association for the Advancement of Science (AAAS). 2007. *Atlas of science literacy.* Vol. 2. Washington, DC: AAAS.

American Association for the Advancement of Science (AAAS). 2009. Benchmarks for science literacy online. *www.project2061.org/publications/bsl/online.*

Driver, R., A. Squires, P. Rushworth, and V. Wood-Robinson. 1994. *Making sense of secondary science: Research into children's ideas.* London: RoutledgeFalmer.

Keeley, P. 2008. *Science formative assessment: 75 strategies for linking assessment, instruction, and learning.* Thousand Oaks, CA: Corwin Press.

Keeley, P. 2011a. Formative assessment probes: Teachers as classroom researchers. *Science and Children* 49 (3): 24–26.

Keeley, P. 2011b. *Uncovering student ideas in life science, vol. 1: 25 more formative assessment probes.* Arlington, VA: NSTA Press.

Keeley, P. 2013. *Uncovering student ideas in primary science, vol. 1: 25 new formative assessment probes for grades K–2.* Arlington, VA: NSTA Press.

Keeley, P. 2014a. *Science formative assessment: 50 more practical strategies for linking assessment, instruction, and learning.* Vol. 2. Thousand Oaks, CA: Corwin Press.

Keeley, P. 2014b. *What are they thinking? Promoting elementary learning through formative assessment.* Arlington, VA: NSTA Press.

Keeley, P. 2015. *Science formative assessment: 75 strategies for linking assessment, instruction, and learning.* 2nd ed. Thousand Oaks, CA: Corwin Press.

Keeley, P., F. Eberle, and C. Dorsey. 2008. *Uncovering student ideas in science, vol. 3: Another 25 formative assessment probes.* Arlington, VA: NSTA Press.

Keeley, P., F. Eberle, and L. Farrin. 2005. *Uncovering student ideas in science, vol. 1: 25 formative assessment probes.* Arlington, VA: NSTA Press.

Keeley, P., F. Eberle, and J. Tugel. 2007. *Uncovering student ideas in science, vol. 2: 25 more formative assessment probes.* Arlington, VA: NSTA Press.

Keeley, P., and C. Sneider. 2012. *Uncovering student ideas in astronomy: 45 new formative assessment probes.* Arlington, VA: NSTA Press.

Keeley, P., and J, Tugel. 2009. *Uncovering student ideas in science, vol. 4: 25 new formative assessment probes.* Arlington, VA: NSTA Press.

Konicek-Moran, R., and P. Keeley. 2015. *Teaching for conceptual understanding in science.* Arlington, VA: NSTA Press.

National Research Council (NRC). 2012. *A framework for K–12 science education: Practices, crosscutting concepts, and core ideas.* Washington, DC: National Academies Press.

NGSS Lead States. 2013. *Next Generation Science Standards: For states, by states.* Washington, DC: National Academies Press. *www.nextgenscience. org/next-generation-science-standards.*

Acknowledgments

We would like to thank the teachers and science coordinators we have worked with for their willingness to field test probes, provide feedback on the format and structure of the probes, share student data, and contribute ideas for assessment probe development. We would especially like to thank Linda Olliver, the extraordinarily talented artist who creatively transforms our ideas into the visual representations seen on the student pages. And of course, our deepest appreciation goes to Claire Reinburg and all the dedicated staff members at NSTA Press who continue to support formative assessment and publish the best books in K–12 science education.

. . . .

Dedications

Page's Dedication

I dedicate this book to Christopher Keeley. I am so proud of all the work you do at the University of New Hampshire Sea Grant program to help communities understand and adapt to climate change. I also dedicate this book to Christine Anderson-Morehouse, Jean-May Brett, and Margo Murphy—three long-time friends, colleagues, extraordinary science education leaders, and passionate environmentally concerned citizens in Maine and Louisiana who have worked with and supported me in uncovering student ideas for almost two decades. I am so proud of all the work you continue to do to support students and teachers and help others appreciate and protect the pristine beauty of my former home state of Maine and the Louisiana wetlands.

Laura's Dedication

This book is dedicated to my husband, Hank—my dear friend since 1970, who is a never-ending source of support and encouragement through all life's challenges, including book deadlines. I also dedicate this body of work to the staff and students of Exploring New Horizons and Port Townsend's Students for Sustainability for bringing such meaning to my life, enriching my soul, and serving as a source of inspiration for me every day.

. . . .

About the Authors

 **Page Keeley** recently retired from the Maine Mathematics and Science Alliance (MMSA) where she was the senior science program director for 16 years, directing projects and developing resources in the areas of leadership, professional development, linking standards and research on learning, formative assessment, and mentoring and coaching. She has been the principal investigator and project director of three National Science Foundation (NSF)–funded projects, including the Northern New England Co-Mentoring Network, PRISMS (Phenomena and Representations for Instruction of Science in Middle School), and Curriculum Topic Study—A Systematic Approach to Utilizing National Standards and Cognitive Research. In addition to NSF-funded national projects, she has designed and directed several state projects, including TIES K–12: Teachers Integrating Engineering into Science K12 and a National Semiconductor Foundation grant project called L-SILL (Linking Science, Inquiry, and Language Literacy). She also founded and directed four cohorts of the Maine Governor's Academy for Science and Mathematics Education Leadership, which is a replica of the National Academy for Science and Mathematics Education Leadership of which she is a Cohort 1 Fellow.

Page is the author of 18 national best-selling books on formative assessment, teaching for conceptual understanding, and curriculum topic study. Currently, she provides consulting services to school districts and organizations throughout the United States on building teachers' and school districts' capacity to use diagnostic and formative assessment and teach for conceptual understanding. She is a frequent invited speaker on formative assessment and teaching for conceptual change.

Page taught middle and high school science for 15 years before leaving the classroom in 1996. At that time, she was an active teacher leader at the state and national level. She served two terms as president of the Maine Science Teachers Association, was a District II NSTA Director, and served as the 63rd President of NSTA in 2008–2009. She received the Presidential Award for Excellence in Secondary Science Teaching in 1992, the Milken National Distinguished Educator Award in 1993, the AT&T Maine Governor's Fellow in 1994, the National Staff Development Council's (now Learning Forward) Susan Loucks-Horsley Award for Leadership in Science and Mathematics Professional Development in 2009, and the National Science Education Leadership Association's Outstanding Leadership in Science Education Award in 2013. She has served as an adjunct instructor at the University of Maine, was a science literacy leader for the American Association for the Advancement of Science /Project 2061 Professional Development Program, and has served on several national advisory boards. She is a science education delegation leader for the People to People Citizen Ambassador Professional Programs, leading the trip to South Africa in 2009, China in 2010, India in 2012, Cuba in 2014, and Peru in 2015.

Before teaching, she was a research assistant in immunology at the Jackson Laboratory of Mammalian Genetics in Bar Harbor, Maine. She received her BS in life sciences from the University of New Hampshire and her MEd

About the Authors

in secondary science education from the University of Maine. She currently resides in Fort Myers, Florida, where in her spare time she dabbles in nature and food photography, culinary art, and cultural travel.

 **Laura Tucker** has been a science educator for 38 years. Initially studying to be a wildlife biologist, she found her passion in teaching students in the outdoors, founding a nonprofit educational organization in 1979 (Exploring New Horizons). The program was designed to provide a comprehensive outdoor environmental science program for K–8 grade students and a summer camp program for children ranging from age 9 to 18. During her tenure, she helped develop a variety of programs, which combined environmental science curricula (redwood, coastal, and Sierra Nevada natural history and ecology, marine biology, botany, zoology, geology, and astronomy) with music, dance, drama, art, and team building. The programs blended the teaching skills and talents of staff naturalists with classroom teachers to incorporate the outdoor school experience into the classroom. Approximately 60,000 students attended the programs while Laura was the executive director. Exploring New Horizons continues to this day, serving about 6,000 students per year on three campuses in the Santa Cruz Mountains of California.

In 1992, she became the professional development coordinator for Great Explorations in Math and Science (GEMS), a nationally acclaimed resource for activity-based science and mathematics at the Lawrence Hall of Science at the University of California, Berkeley. She worked with a variety of educators, including preservice teachers; classroom teachers; district, regional, and state curriculum coordinators; university faculty members; and nonformal educators from museums, zoos, nature centers, and so on. She was a leader in establishing the GEMS Network, which comprises approximately 72 sites and centers around the United States and 11 at international locations. Laura served as a curriculum developer and reviewer for many of the GEMS publications, including *Aquatic Habitats* (Barrett and Willard 1998), *Dry Ice Investigations* (Barber, Beals, and Bergman 1999), *River Cutters* (Sneider and Barrett 1999), and *Schoolyard Ecology* (Barrett and Willard 2001) teacher guides and the GEMS kits and handbooks for leaders, literature, and assessment.

Laura has been actively involved with NSTA and has presented short courses, preconference symposia and workshops at 22 national conferences and 14 regional conferences, including a NASA/NSTA symposium on "Successful Strategies for Involving Parents in Education." Her engaging workshops have also been featured at numerous other conferences, including at science education association meetings in California and Washington.

Laura has been focusing a great deal of her energy on climate education. In 2012, she was selected as a Climate Reality Project presenter and joined former vice president Al Gore and 1,000 educators from 59 countries for three days of intensive training. She is an NOAA Climate Steward as well as a team member with the Climate Change Environmental Education Project-Based Online Learning Community Alliance in partnership with Cornell University, the North American Association for Environmental Education, and the EECapacity Project. She serves as a mentor with Students for Sustainability, a group from Port Townsend High School that is taking action to mitigate climate change at their school, in their community, in their state, and at the national level. They received the

Environmental Protection Agency's President's Environmental Youth Award for Region 10 in 2013. She serves on the Jefferson County/City of Port Townsend Climate Action Committee and chairs the L2020 Climate Action Outreach Group. She attended the Paris Climate Conference, COP21, in December 2015.

Currently, she wears two hats. She is the waste reduction education coordinator for Jefferson County, Washington, teaching the community to reduce, reuse, and recycle. She is also a consultant, providing custom professional development for formal and informal educational programs in hands-on, inquiry-based environmental and STEM (science, technology, engineering, and mathematics) education.

References

Barber, J., K. Beals, and L. Bergman. 1999. *Dry ice investigations.* Berkeley, CA: Great Explorations in Math and Science.

Barrett, K., and C. Willard. 1998. *Aquatic habitats: Exploring desktop ponds.* Berkeley, CA: Great Explorations in Math and Science.

Barrett, K., and C. Willard. 2001. *Schoolyard ecology.* Berkeley, CA: Great Explorations in Math and Science.

Sneider, C., and K. Barrett. 1999. *River cutters.* Berkeley, CA: Great Explorations in Math and Science.

Introduction

Learning is often "in the head" and an aim of the teacher is to help make this learning visible.

—John Hattie (2011)

This is the 10th book in the *Uncovering Student Ideas in Science* series. Adding the 32 probes in this book to the previous nine volumes in the series, science educators now have a collection of 311 different formative assessment probes covering the nature of science and spanning the disciplines of life, Earth, space, and physical science that can be used with K–12 students as well as with teachers in professional development. Thousands of teachers throughout the United States and worldwide have been using these probes. For many teachers, they have become an integral and regular part of their practice. What makes the probes such valuable resources for science education?

25 Features of *USI* Formative Assessment Probes

Uncovering Student Ideas (*USI*) formative assessment probes are unique, science-specific probing questions that uncover what students really think about core ideas in science. They differ in many ways from typical questions provided in instructional materials. Below are 25 features of *USI* probes that distinguish them from other types of questions and formative assessments:

1. *USI* probes are specifically designed to reveal research-identified commonly held ideas of students and adults.

2. *USI* probes always contain two parts—a selected answer choice with distracters that are based on commonly held ideas and an explanation.

3. *USI* probes include extensive Teacher Notes that explain the scientific idea underlying the probe, link to core disciplinary ideas, summarize the cognitive research, and describe curricular and instructional considerations.

4. *USI* probes are not grade-level specific. They can be used across grade spans because misconceptions often follow students from one grade level to the next.

5. *USI* probes are used throughout a cycle of instruction from the beginning, before students are taught the scientific ideas and concepts, and right up to the end of an instructional unit to reflect on how student thinking has changed.

6. *USI* probes provide insight into what students really think about ideas in science, not what they think the teacher wants to hear or what the answer should be on a test.

7. *USI* probes are usually written in student-friendly language without unfamiliar technical terminology.

8. *USI* probes can be used as written assessments or in talk formats. They

Introduction

can be combined with a variety of formative assessment classroom techniques (FACTs).

9. *USI* probes provide engaging opportunities for students to use the scientific practices of modeling, constructing explanations, and engaging in argument.

10. *USI* probes can be modified and adapted for different grade levels and diverse student populations.

11. *USI* probes uncover teachers' common misconceptions as well as students' misconceptions and alternative ideas.

12. *USI* probes use familiar phenomena or situations to engage and motivate students.

13. *USI* probes are intrinsically interesting. They stimulate intellectual curiosity.

14. *USI* probes are assessments *for* learning, rather than assessments *of* learning, and can even be considered assessments *as* learning. The preposition makes a difference!

15. *USI* probes provide teachers with a treasure trove of information about what their students know, do not know, think they know, or struggle to understand.

16. *USI* probes are transformative in changing the culture of the classroom from one in which only right ideas matter to one in which everyone's initial ideas are important, regardless of whether they are right or wrong.

17. *USI* probes transform teachers' fundamental beliefs about teaching and learning.

18. *USI* probes support metacognition—thinking about one's own thinking. They also create the type of cognitive dissonance that happens during the instruction that follows use of a probe when students realize what they used to think no longer works for them.

Experiencing this type of dissonance is what leads to deeper, enduring learning.

19. *USI* probes use signature formats such as the friendly talk and justified list formats. The friendly talk formats model the talk that should happen in science classrooms and portray the diversity of students in today's classrooms. The justified list formats uncover issues of context and gaps in curriculum and instruction.

20. *USI* probes begin as diagnostic assessments. They become formative when the teacher uses the information to plan, adapt, and modify instruction.

21. *USI* probes can be used as both pre- and post-assessments so that students can reflect on how their thinking has changed and teachers can note evidence of conceptual change.

22. *USI* probes target core concepts and ideas that are in most state standards and the *Next Generation Science Standards* (*NGSS;* NGSS Lead States 2013). They are not designed for assessing trivial facts and vocabulary.

23. *USI* probes can be used to encourage a prediction before launching into an investigation to test one's ideas. When an investigation is not practical, they can be used to launch into obtaining information from text and other sources to support or change one's answer choice.

24. *USI* probes support literacy capacities. They can be used to build and support the use of oral language, the norms of discourse in science, and the important skill of careful listening. The explanation part of each probe supports writing in the content area as students construct written explanations.

25. *USI* probes are excellent tools for professional development. They can be used

with teachers to uncover their ideas and strengthen content knowledge. They can also be used with different professional development protocols for examining student thinking.

Earth and Environmental Science Probes in Practice

What does it look like when teachers at different grade levels use the probes in this book? How do teachers combine a formative assessment probe with a FACT? What is the teacher doing during the formative assessment process? What are the students doing? The hypothetical classroom scenarios discussed in this section will help you answer those questions.

Third-Grade Formative Assessment Snapshot: Soil

Picture a third-grade classroom beginning a sequence of lessons about soil. The school is located in an agricultural community. Learning about soil and its importance as a natural resource is part of the school's place-based curriculum. The teacher, Mr. Ortiz, assumed his students already knew a lot about characteristics of soil from their experience living on or near a farm. He selected the justified list formative assessment probe "What Do You Know About Soil?" to elicit students' prior knowledge about their community's important natural resource. Justified list probes use a format that consists of a list of statements, descriptions, or examples of a concept or principle. In this justified list probe, students select the statements that describe soil and explain their reasoning for selecting those statements.

Mr. Ortiz combined the probe with the card sort strategy (Keeley 2008, 2015). In this strategy, the statements (answer choices) for the justified list probe are printed on cards. The students sort the cards into three columns: statements that describe soil, statements that do not describe soil, and statements they are

not sure about yet or do not all agree on. As he circulated among groups, listening to the students discuss where to place their cards, he was quite surprised to hear that most of the students did not think of living and decayed matter as being part of the soil. They thought of it as being in the soil, but separate from it. Almost all of the students referred to dirt and soil synonymously. There were several arguments about how deep soil goes. Some students thought soil extended for at least a mile below the surface of Earth and was one of the main layers of Earth. He also noted a few students talking about how soil gets "used up" by plants and made a note to probe further to see if they thought it no longer existed once it was "used up." He was careful not to immediately correct their misconceptions but was sure to have the students re-examine them after they had an opportunity to explore and learn key ideas about soil. Mr. Ortiz also noted some correct ideas the students were starting with, such as soil comes in different colors and can hold water. He noted cards that students could not yet place in the two categories so he could probe further to find out why students had difficulty with those statements.

Once the students had sorted the cards into the three columns and discussed their reasons for the card placements, Mr. Ortiz had them take a photo of their card sort with their iPads. He told the groups to save the photo for later when they would go back to their original card sort and have an opportunity to change the card placements after they learned about soil. In the meantime, Mr. Ortiz planned a sequence of lessons so that students would have experiences or obtain information that would lead them to re-examine their initial ideas and change them based on new evidence and information. After students completed a sequence of lessons, they pulled up their photo of the initial card sort on their iPads and discussed which cards they would now

Introduction

move and why. Mr. Ortiz culminated the activity with a whole-class consensus card sort on the Smartboard and a discussion based on evidence and information from their lessons, which resolved their initial misconceptions, solidified initial scientific ideas, and resulted in deeper learning.

Fifth-Grade Formative Assessment Snapshot: Groundwater

Picture a fifth-grade classroom learning about freshwater resources. The school is in a state that recently adopted the *NGSS*. The teacher, Ms. Albers, examined the fifth-grade performance expectation, "Describe and graph the amounts and percentages of water and freshwater in various reservoirs to provide evidence about the distribution of water on Earth" (NGSS Lead States 2013). The assessment boundary limited sources of water to the ocean, lakes, rivers, glaciers, groundwater, and polar ice caps. She also examined the disciplinary core idea "ESS2.C: Nearly all of Earth's available water is in the ocean. Most fresh water is in glaciers or underground; only a tiny fraction is in streams, lakes, wetlands, and the atmosphere" (NRC 2012). She noticed that groundwater was included in both the performance expectation and the disciplinary core idea. She wondered if her students had any preconceived ideas about groundwater that might affect their learning. She decided to find out before teaching a lesson on groundwater.

She found a formative assessment probe titled "Groundwater." She read the purpose in the Teacher Notes, "The purpose of this assessment probe is to elicit students' ideas about a major freshwater resource, groundwater. The probe is designed to find out how students visualize groundwater." This probe is exactly what she was looking for. She also liked the friendly talk format, which was set in the context of people talking about science and each person having different ideas. Instead of labeling answer choices

A, B, C, D, the answer choices are listed by the name of the person sharing his or her idea. Ms. Albers thought this format was a good way to not only model what happens in the science classroom when students share their thinking and find out that there are many different ideas, but also make it safe for the reluctant student to share his or her thinking because someone else first states the ideas.

She also liked the suggestion of using the annotated drawing technique as a way for students to support their explanation (Keeley 2008, 2015). An annotated drawing is a visual representation of a student's thinking. The act of drawing to explain a concept or phenomenon promotes thinking and awareness of one's own ideas. Ms. Albers is always looking for ways to use models. Having students draw a cross-sectional view of what they think groundwater looks like below Earth's surface supports the scientific practice of developing and using a model.

Ms. Albers read the research summary in the Teacher Notes and learned that many students think groundwater flows in river-like systems underground or is in pond- or lake-like reservoirs. She was curious to find out whether her students had a similar conception that would appear in their answer choices, explanations, and drawings. Because some of her students would likely hold ideas that mirror the research finding, she decided to plan a lesson to have on hand so that students could experience, with the use of a physical model, how groundwater fills in pores, cracks, and spaces between earth material (such as soil, fractured rock, gravel, and sand) and moves slowly through a formation called an aquifer.

Ms. Albers administered the probe to her students and encouraged them to make a drawing to support their answer choices. She then had the students form pairs and share their thinking. After the pairs discussed their ideas using their drawings, Ms. Albers asked for volunteers to share their thinking with the whole class. She set up the document camera so that students could use their drawings to explain their thinking. Ms. Albers recorded the set of class ideas and then launched into an activity that modeled groundwater so that students could see how water filled the tiny spaces between rock and earth material. She then showed students a video that connected the model used in their activity to actual groundwater and aquifers. She pointed out the chart of class ideas and asked students to vote on which idea they now think best describes groundwater.

After the lesson, Ms. Albers gave students an opportunity to revise or refine their original answers to the probe, including their drawings, using a different color pen or pencil. She collected the probes and reviewed the responses that evening for evidence of conceptual change and deeper understanding.

Middle School Formative Assessment Snapshot: Plate Tectonics

Picture an eighth-grade classroom about to begin a unit on Earth history that includes several lessons on plate tectonics. Their teacher, Dr. Kakkar, was aware that students had learned basic ideas about Earth's plates in sixth grade. He decided to use the formative assessment probe, "Describing Earth's Plates," to elicit students' prior knowledge about tectonic plates. He then decided to use this information to plan his instruction while simultaneously piquing students' interest in the topic and promoting thinking.

Dr. Kakkar read the related research summaries in the Teacher Notes. He was interested in learning more about the Brent and Ford research study on middle school students' ideas about plate tectonics that was cited in the summary (Ford and Taylor 2006). He noticed in the references section that the article was published in the NSTA journal *Science Scope.*

Because Dr. Kakkar is an NSTA member, he had access to all the archived NSTA journals electronically. He found the article and read it. He realized that students' held a myriad of ideas about plate tectonics and Earth's processes. He was particularly struck by the comment that many ideas students have about plate tectonics have been influenced by the language, visual representations, and models that teachers use. Providing an opportunity for students to surface and express their ideas would help him focus on planning instruction that would build a bridge between where his students are in their thinking and where they need to be in order to have a conceptual understanding of plate tectonics and Earth's processes.

Dr. Kakkar decided to use the claims card strategy as a way for students to make their thinking visible (Keeley 2008, 2015). He liked this strategy because it gave students an opportunity to use the scientific practice of constructing an explanation and using it to engage in argumentation with their peers. He printed onto cards each of the 15 statements that made up the answer choices on the probe. He distributed a single card to each student. Because he had 18 students in his class, he made up three additional cards. Each student read the statement on his or her card as a claim. The students then had to support or refute the claim using scientific reasoning supported by evidence. After each student shared and defended or refuted his or her claim, other students could also add ideas to support or refute the claim. After the students finished discussing each claim, the class voted on whether to accept or reject each claim based on the evidence and the strength of the arguments. Dr. Kakkar then put the claims on a chart and sorted them by two columns titled "Initial Claims We Support" and "Initial Claims We Reject." Without passing judgment on students' initial ideas,

Introduction

Dr. Kakkar carefully listened and made note of ideas to address in subsequent instruction.

For example, one student made the claim, "Earthquakes and volcanoes can be found along plate boundaries." She explained the evidence for the claim by sharing what she previously learned about the Pacific Rim of Fire. Other students added ideas such as the number of earthquakes in California and even mentioned the San Andreas Fault. One student added that earthquakes and volcanoes follow the pattern of the plate boundaries, but they can also be found in other areas. A student mentioned how she used to live in Missouri where there was once a powerful earthquake back in the 1800s. When they voted, the entire class agreed to put that event in the "Initial Claims We Support" column.

Another student shared the claim, "Tectonic plates float on top of hot, liquid magma." He supported the claim by explaining that most of Earth's interior is molten, and the plates rest on top of the molten material. He even drew a picture of a cross-section of Earth to support his thinking. Another student agreed and said there are huge pools of magma right under the plates and sometimes it comes up through volcanoes. Another student said that because the plates do not float on water, they must float on magma. After several students agreed with the claim, one student spoke up and said he did not agree with the claim. The plates sit on rock that is sort of melted because it is softened, but it is not a liquid like magma that is melted rock. The material under plates is still considered solid rock even though it is not as hard as the plates that sit on it. The class respectfully considered his argument but voted in favor of the claim that the plates float on top of hot, liquid magma. Dr. Kakkar noted that this claim is a common misconception that he will need to address in subsequent instruction. He knew that merely correcting students would probably not change this strongly held idea. He will plan an opportunity for this idea to surface again when students have more information and recognize the discrepancy between their initial ideas and the scientific ideas they are developing.

After the class presented each claim and their arguments in favor of acceptance or rejection and recorded the claims on the claims chart, Dr. Kakkar had students form small groups to research three or four assigned claims and prepare a new argument, with scientific evidence to support or reject the claim. This constructivist approach provided an opportunity for students to "own" the content, rather than merely filling their heads with information or experiencing activities that do little to confront students' commonly held ideas.

Dr. Kakkar carefully read the Ford and Taylor article again, making note of their suggested strategies for moving student thinking forward as well as the suggestions for instruction and further assessment in the probe Teacher Notes. He kept those suggestions in mind to use while the groups worked on revising or refining their claims.

High School Formative Assessment Snapshot: Climate Change

Picture a high school classroom about to explore the integrated topic of climate change. Their teacher, Mrs. Arquette, wants to activate students' interest with a question that would spark their thinking and uncover any basic misconceptions that would need to be addressed before students learn about the foundation of Earth's global climate system and the evidence that supports a changing climate. Although the difference between weather and climate seems like a basic idea every high school student would know and understand by the end of middle school, Mrs. Arquette thought it would be a good idea to elicit this precursor idea before planning her instruction. She often finds that her high school students hold on to early-formed misconceptions about basic ideas that are part of the elementary and middle school standards. Just because something was taught and assessed several grades before, doesn't mean students learned the content. Often they memorized facts for a test, then completely forgot them and reverted to their strongly held misconceptions.

Mrs. Arquette decided to use the formative assessment probe "Coldest Winter Ever." As she perused the Teacher Notes, she noticed that the notes for administering the probe suggested using the lines-of-agreement strategy (Keeley 2014a). This formative assessment strategy works best with opposing views probes during which two people have opposite or contrasting ideas (in this probe, the two people were Mr. and Mrs. Martin). It also provides an opportunity for students to use the scientific practice of argumentation.

Mrs. Arquette projected the probe onto a screen and had each student complete an individual "quick write" in response to the probe. She then had each student discuss his or her answer with another student. This discussion prepared students for the lines-of-agreement strategy by giving them an opportunity to talk through their ideas first with one other person. Mrs. Arquette found that talking through initial ideas helps students make their thinking visible to others and helps formulate their arguments.

After students discussed their answers in pairs, Mrs. Arquette asked students to form two lines. The students who chose Mr. Martin as their answer choice lined up on one side of the room. The students who chose Mrs. Martin lined up on the other side of the room. The two lines faced each other. Although students were familiar with this FACT, Mrs. Arquette reminded them of the norms they needed to pay attention to during a lines-of-agreement argumentation activity. She also reminded them that if at any time an argument was compelling enough to change their thinking, they could cross over to the other line. She explained how argumentation involves both presenting an evidence-based argument and listening carefully to evaluate the arguments of others for their scientific validity. She told them she would be the facilitator and that she would not comment on their ideas or pass any judgment as to whether they were right or wrong. She wanted all students to freely share their ideas, and she would listen carefully and use their arguments to plan her next steps for instruction. The following is a partial transcript of the argumentation session:

Mrs. Arquette: "Who do you agree with and why? Who would like to start us off?"

Randy: "I agree with Mr. Martin. It's a big change in Florida's climate. Florida is usually hot and sunny, and it doesn't snow there. I think this shows that the Florida climate is changing."

Kaila: "I agree with Randy. It's such a big change from the usual weather they have that I think it's a big change in their climate."

[One student crosses over to the change-in-climate line.]

Introduction

Orlando: "Well, I disagree. I think it's a change in the weather. It's not climate because this is the very kind of thing that proves climate change is not real. If climate change means global warming, then it can't snow or be colder in Florida."

[Two students cross over to the change-in-weather line.]

Lani: "I agree with Orlando that it's about weather, but I don't agree for the same reason. This is just the weather on those days."

Pierce: "Lani makes a good point. There needs to be a pattern of cold winters every year to be a change in climate. This is just a one time thing."

Abdul: "I think it has to be something like 30 years of having a cold winter in Florida to call it a change in climate."

Petra: "Even though it's weather and doesn't happen every year, it can still be related to climate change because a lot of these weird disturbances we hear about are part of the effect of climate change."

Glade: "Yeah, like warm winters up here in Massachusetts. My Dad even played golf once in January."

Randy: "I'm switching what I said before. I think this is a change in the weather, even though it is a big change. Sometimes weather can be pretty extreme, but it is a short-term thing. There isn't enough information here to say it is a change in climate."

[Six students cross over to the change-in-weather line.]

Mrs. Arquette: "Here are some ideas I heard so far about climate. I am not saying these are right or wrong, this is just what you said about changes in climate: it has to be a big change, there needs to be a pattern, it takes 30 years or more to define climate, extreme weather can be related to climate change, cold winters in Florida disprove the idea that Earth is warming, and we need more information to determine change in climate. And here are some of the things I heard you say about changes in weather: weather is a one-time or short-term event, weather can be related to climate change, and weather can involve extreme changes that are related to climate change. Now that we have shared our ideas about weather and climate, let's look at how scientists describe the difference between weather and climate."

Mrs. Arquette shows students a video of a scientist describing the difference between weather and climate and how extreme weather may sometimes be related to a changing climate. As a three-minute exit ticket before leaving class, Mrs. Arquette projects the probe once again and asks students to pick who they most agree with now and explain why, based on their argumentation session and the information in the video. As students leave, they turn in their exit ticket, which Mrs. Arquette will be able to quickly scan to get a picture of where the class is now in their thinking about the difference between weather and climate and the relationship between extreme or unusual weather events and climate change.

Summary of Snapshots

The most important lesson from each of these hypothetical snapshots is that the teacher is using the information about students' thinking to feed back into his or her instruction. Always remember that assessment is not formative unless you use the information to inform your teaching and support learning. If all you do is use a probe to uncover misconceptions and then continue to teach the way you always do or had planned without taking into account students' ideas, you are not practicing formative assessment. Furthermore, as an assessment for learning, you must provide an opportunity for students to revisit their initial ideas and reformulate them based on new evidence and information. We hope the probes in this book will make a difference in your teaching and in your students' learning of important ideas in Earth and environmental science.

Resources for More Information About Formative Assessment Probes

If formative assessment probes are new to you, take some time to acquire the background knowledge in formative assessment you will need to use them effectively. Each of the volumes in the *Uncovering Student Ideas in Science* series includes an introductory chapter that introduces an aspect of formative assessment. Those introductory chapters can be downloaded for free from the NSTA Press website at *www.nsta.org/publications/press*. In addition, the following resources, available though NSTA Press, are widely used as companions to the *Uncovering Student Ideas in Science* series:

- *Science Formative Assessment: 75 Practical Strategies for Linking Assessment, Instruction, and Learning* (2nd edition) (Keeley 2015)
- *Science Formative Assessment, Volume 2: 50 More Strategies for Linking Assessment, Instruction, and Learning* (Keeley 2014a)
- *Teaching for Conceptual Understanding in Science* (Konicek-Moran and Keeley 2015)
- *What Are They Thinking? Promoting Elementary Learning Through Formative Assessment* (Keeley 2014b)

Be sure to visit the *Uncovering Student Ideas in Science* website at *www.uncoveringstudentideas.org* For more information about the series and professional development in formative assessment, contact Page Keeley at *pagekeeley@gmail.com*. You can also follow Page's formative assessment work on Twitter at *@CTSKeeley*.

Introduction

References

Ford, B., and M. Taylor. 2006. Investigating students' ideas about plate tectonics. *Science Scope* 30 (1): 38–43.

Hattie, J. 2011. *Visible learning for teachers: Maximizing impact on learning.* London: Routledge.

Keeley, P. 2008. *Science formative assessment: 75 strategies for linking assessment, instruction, and learning.* Thousand Oaks, CA: Corwin Press.

Keeley, P. 2014a. *Science formative assessment: 50 more practical strategies for linking assessment, instruction, and learning.* Vol. 2. Thousand Oaks, CA: Corwin Press.

Keeley, P. 2014b. *What are they thinking? Promoting elementary learning through formative assessment.* Arlington, VA: NSTA Press.

Keeley, P. 2015. *Science formative assessment: 75 strategies for linking assessment, instruction, and learning.* 2nd ed. Thousand Oaks, CA: Corwin Press.

Konicek-Moran, R., and P. Keeley. 2015. *Teaching for conceptual understanding in science.* Arlington, VA: NSTA Press.

National Research Council (NRC). 2012. *A framework for K–12 science education: Practices, crosscutting concepts, and core ideas.* Washington, DC: National Academies Press.

NGSS Lead States. 2013. *Next Generation Science Standards: For states, by states.* Washington, DC: National Academies Press. *www.nextgenscience.org/next-generation-science-standards.*

Section 1
Land and Water

Concept Matrix.. 12

Related *Next Generation Science Standards* Performance Expectations ... 13

Related NSTA Resources 13

1 **What's Beneath Us?** 15
2 **What Do You Know About Soil?** 19
3 **Land or Water?** 25
4 **Where Is Most of the Fresh Water?** 29
5 **Groundwater** 33
6 **How Many Oceans and Seas?** 37
7 **Why Is the Ocean Salty?** 41

Concept Matrix
Probes #1–#7

Land and Water

CORE CONCEPTS ↓ / PROBES	#1 What's Beneath Us?	#2 What Do You Know About Soil?	#3 Land or Water?	#4 Where Is Most of the Fresh Water?	#5 Groundwater	#6 How Many Oceans and Seas?	#7 Why Is the Ocean Salty?
GRADE RANGE →	3–8	3–8	3–8	3–12	3–12	6–12	5–12
aquifer					X		
concept of one ocean						X	
Earth's crust	X	X					
erosion							X
fresh water				X	X		
global ocean conveyer belt						X	
groundwater				X	X		
hydrosphere			X				
layers of Earth	X						
ocean				X		X	X
ocean basins						X	
rock	X						
salinity							X
salt water							X
seas						X	
soil	X	X					
structure of Earth			X				
water circulation						X	
water distribution			X	X			
weathering							X

Related *Next Generation Science Standards* Performance Expectations (NGSS Lead States 2013)

Earth's Systems

- Grade 2, 2-ESS2-2: Develop a model to represent the shapes and kinds of land and bodies of water in an area.
- Grade 2, 2-ESS2-3: Obtain information to identify where water is found on Earth and that it can be solid or liquid.
- Grade 4, 4-ESS2-2: Analyze and interpret data from maps to describe patterns of Earth's features.
- Grade 5, 5-ESS2-1: Develop a model using an example to describe ways the geosphere, biosphere, hydrosphere, and/or atmosphere interact.
- Grade 5, 5-ESS2-2: Describe and graph the amounts and percentages of water and fresh water in various reservoirs to provide evidence about the distribution of water on Earth.
- Grades 6–8, MS-ESS2-4: Develop a model to describe the cycling of water through Earth's systems driven by energy from the sun and the force of gravity.
- Grades 9–12, HS-ESS2-5: Plan and conduct an investigation of the properties of water and its effects on Earth materials and surface processes.

Reference

NGSS Lead States. 2013. *Next Generation Science Standards: For states, by states.* Washington, DC: National Academies Press. *www.nextgenscience.org/next-generation-science-standards.*

Related NSTA Resources
NSTA Press Books

Aretxabaleta, A., G. Brooks, and N. West. 2011. *Project Earth science: Physical oceanography.* 2nd ed. Arlington, VA: NSTA Press.

Blake, R., J. A. Frederick, S. Haines, and S. C. Lee. 2010. *Inside-out: Environmental science in the classroom and the field, grades 3–8.* Arlington, VA: NSTA Press.

Konicek-Moran, R. 2013. *Everyday Earth and space science mysteries: Stories for inquiry-based science teaching.* Arlington, VA: NSTA Press.

Rich, S. 2012. *Bringing outdoor science in: Thrifty classroom lessons.* Arlington, VA: NSTA Press.

U.S. Department of Agriculture Natural Resources Conservation Service, and National Science Teachers Association. 2001. *Dig in! Hands-on soil investigations.* Arlington, VA: NSTA Press.

NSTA Journal Articles

Beckrich, A. 2014. The green room: Get grounded in groundwater. *The Science Teacher* 81 (2): 8.

Clary, R. 2015. The resource beneath our feet. *The Science Teacher* 82 (6): 49–56.

Colburn, A. 2010. The prepared practitioner: Groundwater—a green issue. *The Science Teacher* 77 (2): 8.

Cronin, J. 2004. Great globes. *Science Scope* 28 (2): 26–27.

Keeley, P. 2015. Soil and dirt: The same or different? *Science and Children* 52 (7): 29–31.

Leuenberger, T., D. Shepardson, J. Harbor, C. Bell, J. Meyer, H. Klagges, and W. Burgess. 2001. Inquiry and aquifers. *Science Scope* 25 (2): 20–26.

Mayes, V. 2009. Natural resources: Digging soil. *Science and Children* 47 (3): 44–45.

Robertson, B. 2009. Science 101: Why are oceans salty and lakes and rivers not? *Science and Children* 46 (8): 62–63.

Schoedinger, S., F. Cava, and B. Jewell. 2006. The need for ocean literacy in the classroom. *The Science Teacher* 73 (6): 44–52.

Taylor, C., and C. Graves. 2010. Soil is more than just dirt. *Science Scope* 33 (8): 70–75.

NSTA Learning Center Resources
NSTA Webinars
List of NOAA Ocean–Related Webinars *http://learningcenter.nsta.org/products/web_seminar_archive_sponsor.aspx?page=NOAA*

NGSS Core Ideas: Earth's Systems
http://learningcenter.nsta.org/products/
symposia_seminars/NGSS/webseminar32.aspx

NSTA Science Objects
Earth, Moon, and Sun: General Characteristics of Earth
http://learningcenter.nsta.org/product_detail.
aspx?id=10.2505/7/SCB-ESM.1.1

What's Beneath Us?

Five friends were digging in their garden. They each had different ideas about what Earth would be like if they could dig deep down for about 10 miles. This is what they said:

Haley: I think it would be mostly solid rock with a very thin layer of soil on top.

Ariel: I think it would be mostly big rocks, small stones, and gravel, with a layer of soil on top.

Kip: I think it would be mostly soil with a lot of rocks scattered throughout the soil.

Tyler: I think it would be mostly big rocks with a thin layer of soil on top and soil filling in the spaces between the rocks.

Frieda: I think there would be repeating layers of soil and rock. With each layer, the rocks under the soil get bigger.

Which friend do you think has the best idea? _____ Explain your thinking.

_____

_____

_____

_____

What's Beneath Us?

Teacher Notes

Purpose

The purpose of this assessment probe is to elicit students' ideas about the structure of the solid Earth. The probe is designed to find out if students recognize that Earth's outer layer (crust) is mostly solid rock with a thin layer of soil on top.

Type of Probe

Friendly talk

Related Concepts

Earth's crust, layers of Earth, rock, soil

Explanation

The best answer is Haley's: "I think it would be mostly solid rock with a very thin layer of soil on top." Soil covers much of the surface of Earth in a very thin layer. This soil is formed very slowly as rock weathers and tiny pieces mix with organic and inorganic material. Soil depth varies greatly by location, but the average worldwide depth is about 15 cm (6 in.). Although the depth of the crust varies, if you were on land and could dig down to about 10 miles, Earth would be mostly rock, with loose materials at the surface such as soil and pieces of rock. This part near the surface of Earth is called the crust. As you dig deeper beyond the *crust,* you get to the mantle. The upper part of the mantle is made up of solid rock. The crust and the upper part of the mantle make up the solid, outer layer of Earth, which is called the *lithosphere.* The lithosphere is divided into tectonic plates.

Administering the Probe

This probe is best used with grades 3–8. Extend the probe by encouraging students to draw a picture to visually represent their conceptual model. This probe can be combined with the formative assessment classroom technique, annotated drawings (see pp. 4–5).

Related Core Ideas in *Benchmarks for Science Literacy* (AAAS 2009)

3–5 Processes That Shape the Earth

- Smaller rocks come from the breakage and weathering of bedrock and larger rocks. Soil is made partly from weathered rock, partly from plant remains—and also contains many living organisms.

6–8 Processes That Shape the Earth

- Although weathered rock is the basic component of soil, the composition and texture of soil and its fertility and resistance to erosion are greatly influenced by plant roots and debris, bacteria, fungi, worms, insects, rodents, and other organisms.

Related Core Ideas in *A Framework for K–12 Science Education* (NRC 2012)

3–5 ESS2.A: Earth Materials and Systems

- Earth's major systems are the geosphere (solid and molten rock, soil, and sediments), the hydrosphere (water and ice), the atmosphere (air), and the biosphere (living things, including humans). These systems interact in multiple ways to affect Earth's surface materials and processes.

Related Research

- When Happs (1984) asked children how far down they thought soil would go if they dug down in their garden, their responses ranged from 6 in. to 10 miles. Some students suggested that soil would extend all the way to Earth's core.
- Children's ideas about the Earth beneath their feet are influenced by their everyday experience digging holes. Few children have experienced digging holes greater than a few feet. When they have dug holes in the Earth or observed excavations, most

children have seen soil with rocks in it (Russell et al. 1993).

Suggestions for Instruction and Assessment

- This probe is best used as a precursor to elicit ideas related to several of the standards and core ideas that involve weathering and erosion, Earth's geological history, and plate tectonics. Students first need to understand that the next layer of Earth beneath them, after the crust, is made up of solid, rigid rock.
- This probe provides an opportunity for students to use the scientific practice of developing and using a model. They can draw a picture to represent their conceptual model and use it to support their explanation. Have students present their models and clarify their ideas about depth and materials. For the deepest part of their drawing, ask students if there is anything under that.
- Be aware that students' experience digging into the ground may affect their thinking because they see mostly soil and pieces of rock.
- Use accurate diagrams and representations to help students visualize Earth's interior structure.
- Observing or showing photos of a steep rock cliff, road cut, or rock quarries will help students see that there is a thin layer of soil and rock rubble on top of a deep layer of solid rock.
- Make sure students understand that soil depth varies and so do other characteristics of soil. In some areas, soil may extend deeply, whereas in other areas, it is quite shallow.

References

American Association for the Advancement of Science (AAAS). 2009. Benchmarks for science literacy online. *www.project2061.org/publications/bsl/online.*

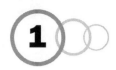

Happs, J. 1984. Soil genesis and development: Views held by New Zealand students. *Journal of Geography* 83 (4): l77–180.

National Research Council (NRC). 2012. *A framework for K–12 science education: Practices, crosscutting concepts, and core ideas.* Washington, DC: National Academies Press.

Russell, T., D. Bell, K. Longden, and L. McGuigan. 1993. *Rocks, soil and weather.* Liverpool, U.K.: Liverpool University Press.

What Do You Know About Soil?

Have you ever picked up a handful of soil? How would you describe it? Put an X next to any statements that generally describe soil.

_____ **A.** Soil comes in different colors.

_____ **B.** Soil is the same as dirt.

_____ **C.** Soil starts as clay and then hardens to form rock.

_____ **D.** Soil makes up most of Earth's crust.

_____ **E.** Soil comes out of volcanoes.

_____ **F.** Soil contains water.

_____ **G.** Soil was present when Earth first formed.

_____ **H.** Soil changes over time.

_____ **I.** Living things are in soil but are not part of the soil.

_____ **J.** Soil depth ranges from a few inches to a mile below the surface.

_____ **K.** Soils are eventually used up by plants and extensive farming.

_____ **L.** Soils can form in deserts.

_____ **M.** Air is a part of soil.

_____ **N.** Rotted material fertilizes the soil but does not become part of the soil.

_____ **O.** Soil is a natural resource.

Explain your thinking. Describe what you know about soil.

_____

_____

_____

_____

What Do You Know About Soil?

Teacher Notes

Purpose

The purpose of this assessment probe is to elicit students' ideas about soil. The probe is designed to uncover common misconceptions about characteristics of soil and its formation.

Type of Probe

Justified list

Related Concepts

Earth's crust, soil

Explanation

The best answers are: A, F, H, L, M, and O. Soil is one of Earth's natural resources and it differs in color, texture, and composition. It is a mixture found on all land masses and contains inorganic material such as minerals, water, and air. Soil also consists of organic material (such as decayed plant and animal life), fungi, small animals (such as insects and mites), and micro-organisms. Soil is formed from the interaction of several environmental factors, including climate, living things, atmosphere, topography, and parent material, and it can range in age from newly formed soil to millions of years old. Soils are dynamic and constantly evolving because of the interaction of environmental factors.

Soil and dirt are often used synonymously, but they are not the same thing. *Soil* is the material that is useful to plants and animals, including people. *Dirt* is soil and other debris in places where humans do not want it, such as dirt on your floor or clothes. Soil forms from the weathering of rock and biological processes. Some of the weathered rock further breaks down in soil, forming finer particles that become clay or silt. However, this formation is not part of a soil to rock process.

Soil makes up only a very thin veneer on Earth's crust. Soil depth varies across the world, with the average depth being about 15 cm (6 in.) Soil does not come out of volcanoes, but the rock that forms from the lava can weather and become part of the soil. Soil has not been

around as long as Earth has existed. Soil did not form until rocks formed, and these rocks weathered over time and combined with other material to begin forming early soils. Soil formation is continuous and still happening today. Soil formation is a slow process. The time it takes to form an inch of soil varies. It is commonly stated that it takes about 100 years to form 1 in. of soil; yet rates of 1 in. of growth in 1,000 years are not uncommon for harder rocks, and some types of soil take even longer. Soil is not used up by plants and farming. Nutrients found in the soil may be removed and used by plants, but the soil itself does not disappear.

Administering the Probe

This probe is best used with grades 3–8. Before using the probe, go over each of the answer choices to make sure students understand the statement, or eliminate statements that may be unfamiliar to your students. This probe can be used with the card sort formative assessment classroom technique (see pp. 3–4).

Related Core Ideas in *Benchmarks for Science Literacy* (AAAS 2009)

. . . .

3–5 Processes That Shape the Earth

- Rock is composed of different combinations of minerals. Smaller rocks come from the breakage and weathering of bedrock and larger rocks. Soil is made partly from weathered rock, partly from plant remains—and also contains many living organisms.

6–8 Processes That Shape the Earth

- Although weathered rock is the basic component of soil, the composition and texture of soil and its fertility and resistance to erosion are greatly influenced by plant roots and debris, bacteria, fungi, worms, insects, rodents, and other organisms.

Related Core Ideas in *A Framework for K–12 Science Education* (NRC 2012)

. . . .

3–5 ESS2.A: Earth Materials and Systems

- Earth's major systems are the geosphere (solid and molten rock, soil, and sediments), the hydrosphere (water and ice), the atmosphere (air), and the biosphere (living things, including humans). These systems interact in multiple ways to affect Earth's surface materials and processes.

6–8 ESS2.A: Earth Materials and Systems

- The planet's systems interact over scales that range from microscopic to global in size, and they operate over fractions of a second to billions of years. These interactions have shaped Earth's history and will determine its future.

Related Research

- Much of the research on children's ideas about soil was conducted in New Zealand in the 1980s. Many New Zealand students believed soil was formed at the same time Earth was formed and has always been here. Even older students failed to explain how soil changes over time. When students were asked what soil is, they tended to use the terms *dirt* and *soil* synonymously. Many of the students in the study believed that soil goes through changes where it eventually becomes clay and then compresses into rock. When asked how deep soil goes, most students thought soil extended several kilometers into Earth (Happs 1982).
- Student interviews about the "disappearance" of dead animals or fruits on the surface of the soil revealed several misconceptions about decay. Young children think dead things just disappear and no longer exist, not accounting for them becoming part of the soil. Some students recognize that rotted material "fertilizes" the soil but do

not recognize it as being part of the soil. Generally, children are unaware of the role micro-organisms play in soil (Driver et al. 1994).

- A study by Gosselin and Macklem-Hurst (2002) found that many preservice teachers believe soil is put down as rock layers, which may be the result of their focusing on horizons in a soil profile.

Suggestions for Instruction and Assessment

- This probe can be used as a card sort strategy (Keeley 2015). Place the answer choices on cards and have students work in small groups to sort them into statements that describe soil and statements that do not accurately describe soil. Encourage students to discuss and defend their reasons for their card placement.
- Provide samples of soil or have students bring them in from their local areas. Have students examine the soil samples with hand lenses and describe them in terms of particle size, color, texture, organic matter, and so on. They can conduct a soil profile by putting 1 in. of soil in a clear 12 oz. cup. Have them fill the cup with water and stir vigorously. The larger particles in the soil will settle to the bottom, with the successively smaller particles forming layers (gravel, sand, silt, clay). Organic material will float at the top. (Safety notes: Make sure soil samples are from pesticide- and herbicide-free sources. Caution students about possible sharp materials in the soil that could cut skin. Immediately wipe up any spilled water to avoid slips and falls. Be sure students wash their hands with soap and water after handling soil.)
- Compare and discuss the difference between naturally formed soil and potting soil.

- Have students explore the process of soil formation by researching how soil appears on a newly formed, barren volcanic island.
- Using an apple as a model, cut the apple in half and have students look at the depth of the peel. Tell students the peel represents Earth's crust and that the layer of soil on top of the crust would be just a very tiny fraction of the thickness of the peel.
- Help students understand the difference between dirt and soil. At a special Smithsonian exhibit on soils, a number of speakers at the media preview of the soil exhibit acknowledged how they initially did not know that soil and dirt were not the same thing. Dirt is an example of an everyday word that is used incorrectly in science. To a soil scientist, dirt and soil are not the same thing. When the curator of an exhibit at the Smithsonian (and a member of the Soil Science Society of America) was asked to explain the difference, he said that dirt is basically soil that has been displaced. In other words, dirt often ends up in places where humans do not want it (Raloff 2008).
- Contact your local soil conservation district and invite one of their scientists to speak to your students.

References

American Association for the Advancement of Science (AAAS). 2009. *Benchmarks for science literacy online. www.project2061.org/publications/bsl/online.*

Driver, R., A. Squires, P. Rushworth, and V. Wood-Robinson. 1994. *Making sense of secondary science: Research into children's ideas.* London: RoutledgeFalmer.

Gosselin, D., and J. Macklem-Hurst. 2002. Pre-/post-knowledge assessment of an Earth science course for elementary/middle school education majors. *Journal of Geoscience Education* 50 (2): 169–175.

Happs, J. 1982. Some aspects of student understanding of soil. *Australian Science Teacher Journal* 28 (3): 25–31.

Keeley, P. 2015. *Science formative assessment: 75 strategies linking assessment, instruction, and learning.* 2nd ed. Thousand Oaks, CA: Corwin Press.

National Research Council (NRC). 2012. *A framework for K–12 science education: Practices, crosscutting concepts, and core ideas.* Washington, DC: National Academies Press.

Raloff, J. 2008. Dirt is not soil. Science News. *www.sciencenews.org/blog/science-public/dirt-not-soil.*

Land or Water?

Three students were talking about the surface of Earth. They each had different ideas about what covers most of Earth's surface. This is what they said:

Mona: I think most of the surface of Earth is covered by land.

Bodvar: I think most of the surface of Earth is covered by water.

Eliete: I think Earth's surface is about half water and half land.

Who do you agree with the most? _____ Explain why you agree.

_____

_____

_____

_____

_____

_____

_____

_____

Land or Water?

Teacher Notes

Purpose

The purpose of this assessment probe is to elicit students' ideas about the surface of Earth. The probe is designed to find out if students recognize that most of Earth's surface is covered by water.

Type of Probe

Friendly talk

Related Concepts

Hydrosphere, ocean, structure of Earth, water distribution

Explanation

The best answer is Bodvar's: "I think most of the surface of Earth is covered by water." Water covers about 71% of Earth's surface. Of this 71%, 97% is ocean. Fresh water from ice covers composes 2% of Earth's total water, with the rest coming from groundwater, rivers, lakes, and other fresh surface waters. About 29% of Earth's surface is covered by land,

with the Asian continent making up about 30% of Earth's total land surface. Deserts cover about 20% of Earth's land area. About 12% of Earth's total land area is developed land inhabited by humans.

Administering the Probe

This probe is best used with grades 3–8. Probe further by asking students to estimate the percentage of Earth covered by land and by water. Remind students that ice counts as water in the solid form.

Related Core Ideas in *Benchmarks for Science Literacy* (AAAS 2009)

6–8 The Earth

- The Earth is mostly rock. Three-fourths of the Earth's surface is covered by a relatively thin layer of water (some of it frozen), and the entire planet is surrounded by a relatively thin layer of air.

Related Core Ideas in *A Framework for K–12 Science Education* (NRC 2012)

3–5 ESS2.A: Earth Materials and Systems

- Earth's major systems are the geosphere (solid and molten rock, soil, and sediments), the hydrosphere (water and ice), the atmosphere (air), and the biosphere (living things, including humans).

3–5 ESS2.C: The Role of Water in Earth's Surface Processes

- Nearly all of Earth's available water is in the ocean. Most fresh water is in glaciers or underground; only a tiny fraction is in streams, lakes, wetlands, and the atmosphere.

Related Research

- Although the authors found no formal research on students' ideas related to water and land distribution, examining results from the formative assessment probe "Where Would It Fall?" reveals that elementary students commonly think that most of Earth is covered by land (Keeley and Tugel 2009).

- In a study of junior high students' perceptions of the water cycle, students were asked to agree or disagree with the following statement: Most of the water in our planet is the salty water in the ocean and is not available for humans to use. Of those surveyed, 12.6% agreed, 23.6% were uncertain, and 63.8% disagreed (Ben-zvi-Assarf and Orion 2005).

Suggestions for Instruction and Assessment

- Have students examine a globe or map of the world to compare Earth's coverage by land masses with coverage by water and ice.
- Combine this probe with the P-E-O (Predict-Explain-Observe) technique and the scientific practice of using a model

to test ideas. Have students predict the percentage of water and land. Using an inflatable globe "beach ball" as a model, have students toss and catch the ball back and forth 25 times or more, each time recording whether their right index finger touches land or water on the ball when they catch it. Have students tally their responses. Probability indicates that most of their catches will land on ocean. Discuss how the model was used to test their predictions. Older students can compare their results with the actual percentages (71% water and 29% land). They can also discuss whether this activity is an accurate way to measure the amount of land and water on Earth. (Safety note: Conduct this activity in an open area that is free of furniture and fragile materials.) Have middle school students develop a graphical representation to compare the coverage of developed and undeveloped land masses, ice covered surfaces, ocean water, and freshwater bodies.

- After learning about the ratio of water to land on Earth, students can discuss whether our planet should be called Planet Earth or Planet Ocean. They can support their opinions with evidence.

- Observe an image of the "blue ball" Earth from space. Discuss how astronauts would answer the probe after gazing at Earth from space.

References

American Association for the Advancement of Science (AAAS). 2009. Benchmarks for science literacy online. *www.project2061.org/publications/bsl/online*.

Ben-zvi-Assarf, O., and N. Orion. 2005. A study of junior high students' perceptions of the water cycle. *Journal of Geoscience Education* 53 (4): 366–373.

Keeley, P., and J. Tugel. 2009. *Uncovering student ideas in science, vol. 4: 25 new formative assessment probes.* Arlington, VA: NSTA Press.

National Research Council (NRC). 2012. *A framework for K–12 science education: Practices, crosscutting concepts, and core ideas.* Washington, DC: National Academies Press.

Where Is Most of the Fresh Water?

Three friends were talking about fresh water. They each had a different idea about where most of Earth's fresh water is found. This is what they said:

Sephali: I think most of Earth's fresh water is found in snow, ice caps, glaciers, and under the ground.

Mary: I think most of Earth's fresh water is found in lakes, rivers, streams, and swamps.

Pearl: I think most of Earth's fresh water is found in the ocean, seas, and bays.

Who do you agree with the most? _____ Explain your thinking.

_____

_____

_____

_____

_____

Where Is Most of the Fresh Water?

Teacher Notes

Purpose

The purpose of this assessment probe is to elicit students' ideas about the distribution of fresh water. The probe is designed to find out if students recognize that most of Earth's fresh water is found in frozen form or groundwater.

Type of Probe

Friendly talk

Related Concepts

Fresh water, groundwater, water distribution

Explanation

The best answer is Sephali's: "I think most of Earth's fresh water is found in snow, ice caps, glaciers, and under the ground." Most of Earth's water comes from the ocean and is saline. Only about 3% of Earth's water is fresh water. Most fresh water is found in the form of ice, snow, groundwater, and soil moisture. Only about 0.3% is found in liquid form on the surface

in lakes, swamps, rivers, and streams. A small percentage of water on Earth is also found in living things and the atmosphere.

Administering the Probe

This probe is best used with grades 3–12. The probe can be extended by asking students to draw a model showing how they think Earth's fresh water is distributed.

Related Core Ideas in *Benchmarks for Science Literacy* (AAAS 2009)

6–8 The Earth

- The Earth is mostly rock. Three-fourths of the Earth's surface is covered by a relatively thin layer of water (some of it frozen), and the entire planet is surrounded by a relatively thin layer of air.
- Fresh water, limited in supply, is essential for some organisms and industrial processes. Water in rivers, lakes, and underground

can be depleted or polluted, making it unavailable or unsuitable for life.

Related Core Ideas in *A Framework for K–12 Science Education* (NRC 2012)

K–2 ESS2.C: The Roles of Water in Earth's Surface Processes

- Water is found in the ocean, rivers, lakes, and ponds. Water exists as solid ice and in liquid form.

3–5 ESS2.C: The Roles of Water in Earth's Surface Processes

- Nearly all of Earth's available water is in the ocean. Most fresh water is in glaciers or underground; only a tiny fraction is in streams, lakes, wetlands, and the atmosphere.

Related Research

Although the authors found no formal research on students' ideas about water distribution, the author's interviews with children ages 10–12 revealed that many thought that fresh water was found mostly in rivers and lakes.

Suggestions for Instruction and Assessment

- Combine this probe with "Where Would It Fall?"—a probe that targets Earth's total water distribution (Keeley and Tugel 2009).
- The U.S. Geological Survey's Water Science School website has several good representations that show Earth's freshwater distribution. Use those representations to help students visualize where fresh water comes from. This website is available at *http://water.usgs.gov/edu/earthwherewater.html*.
- Have students create a physical model of the distribution of Earth's fresh water and

use their model to explain the critical need for water conservation.

- Use an apple as a model to demonstrate the amount of fresh water on Earth. Cut the apple into four quarters. Because Earth is one-fourth land, remove one quarter to represent the land. Three quarters remain to represent the water that covers Earth. Remove the skin off one of the quarters to represent the volume of the 3% of the water that is fresh, relative to the saltwater quarters. Cut the skin into thirds. Two thirds are set aside to represent the fresh water that is frozen as glaciers and polar ice. The last third of apple skin represents all the fresh water on Earth to be used by plants, animals, and humans. This illustration can lead to a deeper discussion of water resources and the scarcity of clean, fresh drinking water and its global implications.
- NASA has created a short lesson on the distribution of water on Earth with a graphic representation of the specific breakdown of the types of water (ocean, groundwater, glaciers, etc.). You can find this resource at *http://pmm.nasa.gov/education/lesson-plans/freshwater-availability-classroom-activity*.

References

American Association for the Advancement of Science (AAAS). 2009. Benchmarks for science literacy online. *www.project2061.org/publications/bsl/online*.

Keeley, P., and J. Tugel. 2009. *Uncovering student ideas in science, vol. 4: 25 new formative assessment probes*. Arlington, VA: NSTA Press.

National Research Council (NRC). 2012. *A framework for K–12 science education: Practices, crosscutting concepts, and core ideas*. Washington, DC: National Academies Press.

Groundwater

Water found below Earth's surface is called groundwater. Five friends wondered what they would see if they could look underground and see groundwater. This is what they said:

Tyson: I think I would see a moving underground stream or river.

Yalena: I think I would see water in the tiny cracks and spaces between soil, sand, and rocks.

Jake: I think I would see a pool of water, sort of like an underground lake.

Betsy: I think I would see water spouting up from a vent or opening deep under the ground.

Armando: I think I would see chunks of ice that slowly melt and release water.

Who do you agree with the most? _____ Explain your thinking.

_____

_____

_____

_____

Groundwater

Teacher Notes

Purpose

The purpose of this assessment probe is to elicit students' ideas about a major fresh-water resource, groundwater. The probe is designed to find out how students visualize groundwater.

Type of Probe

Friendly talk

Related Concepts

Aquifer, fresh water, groundwater

Explanation

The best answer is Yalena's: "I think I would see water in the tiny cracks and spaces between soil, sand, and rocks." Groundwater is water found below the surface of Earth. It is the major source of water for drinking and agriculture. Groundwater is found in the pores, cracks, and spaces between earth material such as soil, fractured rock, gravel, and sand.

It moves slowly through a formation called an *aquifer*.

Administering the Probe

This probe is best used with upper elementary, middle, and high school students. The probe can be extended by asking students to draw a conceptual model showing what they think groundwater looks like from a cross-sectional view below Earth's surface.

Related Core Ideas in *Benchmarks for Science Literacy* (AAAS 2009)

6–8 The Earth

- Water evaporates from the surface of the Earth, rises and cools, condenses into rain or snow, and falls again to the surface. The water falling on land collects in rivers and lakes, soil, and porous layers of rock, and much of it flows back into the oceans.

Related Core Ideas in *A Framework for K–12 Science Education* (NRC 2012)

3–5 ESS2.C: The Role of Water in Earth's Surface Processes

- Nearly all of Earth's available water is in the ocean. Most fresh water is in glaciers or underground; only a tiny fraction is in streams, lakes, wetlands, and the atmosphere.

Related Research

- A common misconception of both students and teachers is that water under the ground flows in river-like systems or in large underground lake-like reservoirs. Students who think this sometimes assume that wells will provide water forever because they are filled by underground rivers. The scale of the spaces water fills also varies from micro to macro, with some older students thinking that groundwater fills spaces the size and depth of skyscrapers. (Dickerson et al. 2007).

- The common misconception that groundwater is like an underground lake may come from the level of abstraction that is needed to understand hidden phenomena and processes that take place underground. Research indicates that students' mental model of groundwater as a static sub-surface lake results from their actual experience with the upper water system (Ben-zvi-Assarf and Orion 2005).

Suggestions for Instruction and Assessment

- Use a sponge to represent how water falling on the surface of Earth seeps into the ground and fills empty spaces. The sponge and water model the porous rock, soil, and sand and the empty spaces between the earth material. (Safety note: Immediately wipe up any spilled water to avoid slips and falls.)

- Student and teacher information about groundwater can be found on the Groundwater Foundation's website at *http://groundwater. org.*

- Visualizing groundwater can be challenging for students, as it is very different from water resources they see on the surface of Earth. Have students draw a conceptual model of the form and location of groundwater.

- A video of sixth graders discussing their ideas about groundwater (including misconceptions) can be viewed at *http:// education.nationalgeographic.com/media/ what-groundwater.*

- Water cycle diagrams may interfere with students' understanding of groundwater as part of the water cycle because many water cycle diagrams show only surface water. Use water cycle diagrams that also show groundwater or have students create water cycle diagrams that include groundwater as part of the cycle.

- Include the use of rock specimens when teaching about groundwater. Students can examine rocks to observe differences between rock types found in aquifers. (Safety note: Instruct students to handle rock specimens cautiously because some rocks may have sharp edges that can cut skin.)

References

American Association for the Advancement of Science (AAAS). 2009. Benchmarks for science literacy online. *www.project2061.org/ publications/bsl/online.*

Ben-zvi-Assarf, O., and N. Orion. 2005. A study of junior high students' perceptions of the water cycle. *Journal of Geoscience Education* 53 (4): 366–373.

Dickerson, D., J. Penick, K. Dawkins, and M. Van Sickle. 2007. Groundwater in science education. *Journal of Science Teacher Education* 18 (1): 45–61.

National Research Council (NRC). 2012. *A framework for K–12 science education: Practices, crosscutting concepts, and core ideas.* Washington, DC: National Academies Press.

6

How Many Oceans and Seas?

A group of friends went sailing on the ocean. They wondered how many separate oceans and seas there were in the world. This is what they said:

Kendra: I counted the bodies of water on a map. There are more than 100 separate oceans and seas.

Matthias: I think there are seven separate oceans or seas because they have always been called the Seven Seas.

Alejandra: I think there is really only one ocean because water flows freely through all of them.

Tidir: Because they are named for the basins that form them, I count 5 separate oceans and 87 separate seas.

Who do you think has the best idea? _____ Explain your thinking.

_____

_____

_____

_____

How Many Oceans and Seas?

Teacher Notes

Purpose

The purpose of this assessment probe is to elicit students' ideas about the ocean and seas. The probe is designed to reveal whether students use geographic names and locations to consider the number of separate oceans and seas or use the concept of "one ocean."

Type of Probe

Friendly talk

Related Concepts

Concept of one ocean, global ocean conveyer belt, ocean, ocean basins, seas, water circulation

Explanation

The best answer is Alejandra's: "I think there is really only one ocean because water flows through all of them." It is important for students and adults to understand the concept of "one ocean" to understand the global movement of matter and the flow of energy throughout the world's ocean. Throughout the ocean, there

is one interconnected circulation system powered by the wind, tides, the force of Earth's rotation (Coriolis effect), the Sun, and water density differences. The shape of ocean basins and adjacent land masses influence the path of circulation. This "global ocean conveyor belt" moves water throughout all of the ocean's basins, transporting energy (heat), matter, and organisms around the ocean (Ocean Literacy Network 2015).

Although the basins have names that contain the named oceans, all ocean water can move freely around the globe and mix with other named oceans. The term *Seven Seas* has historically meant the Arctic, North Atlantic, South Atlantic, North Pacific, South Pacific, Indian, and Southern Oceans. Currently, a sea is commonly thought to be an extended body of saline water associated with one of the world's five named oceans (Atlantic, Pacific, Arctic, Indian, and Southern). Because there is no strict scientific definition of the term *sea*, it is not surprising that there is no single

defined list of the seas of the world. Some maps may list more than 100 "seas" around the globe.

Administering the Probe

This probe is best used with grades 6–12. Being careful not to give away the answer, point out that the term *separate oceans* means bodies of ocean water that are physically separated from other bodies of ocean water.

Related Core Ideas in *Benchmarks for Science Literacy* (AAAS 2009)

6–8 The Earth

- Three-fourths of the Earth's surface is covered by a relatively thin layer of water (some of it frozen), and the entire planet is surrounded by a relatively thin layer of air.

6–8 Processes That Shape the Earth

- Thermal energy carried by ocean currents has a strong influence on climates around the world.

9–12 Processes That Shape the Earth

- Transfer of thermal energy between the atmosphere and the land or oceans produces temperature gradients in the atmosphere and the oceans. Regions at different temperatures rise or sink or mix, resulting in winds and ocean currents. These winds and ocean currents, which are also affected by the Earth's rotation and the shape of the land, carry thermal energy from warm to cool areas.

Related Core Ideas in *A Framework for K–12 Science Education* (NRC 2012)

6–8 ESS2.A: Earth Materials and Systems

- All Earth processes are the result of energy flowing and matter cycling within and among the planet's systems.

6–8 ESS2.C: The Roles of Water in Earth's Surface Processes

- Global movements of water and its changes in form are propelled by sunlight and gravity.

Related Research

Feller (2007) identified several common misconceptions about the ocean, including the commonly held idea that the three big oceans are not connected and each acts alone.

Suggestions for Instruction and Assessment

- Give an inflatable globe to each pair of students. A flat map of the world that is connected as a cylinder can work as well. Have them put a finger on a part of ocean water. Have them attempt to trace a path around the globe or map without crossing any landforms. This should reinforce the idea that we have only one ocean that is connected to all oceans worldwide.
- Examine the Ocean Literacy Framework at *http://oceanliteracy.wp2.coexploration. org*. The Framework includes the Seven Principles of Ocean Literacy, including Principle #1: The Earth has one big ocean with many features. The Framework also includes a scope and sequence document and graphical conceptual flow maps.
- Seas of the World (Saundry 2013) can be used to support the scientific practice of obtaining, evaluating, and communicating information related to this probe.
- Students can research the history behind the names of the oceans and seas on our maps.

References

American Association for the Advancement of Science (AAAS). 2009. Benchmarks for science literacy online. *www.project2061.org/ publications/bsl/online*.

Feller, R. 2007. 110 misconceptions about the ocean. *Oceanography* 20 (4): 170–173.

National Research Council (NRC). 2012. *A framework for K–12 science education: Practices, crosscutting concepts, and core ideas.* Washington, DC: National Academies Press.

Ocean Literacy Network. 2015. Ocean literacy principle #1. In *Ocean literacy framework*, ed. Ocean Literacy Network. Berkeley, CA: Centers for Ocean Sciences Education Excellence. *http:// oceanliteracy.wp2.coexploration.org.*

Saundry, P. 2013. Seas of the world. *The encyclopedia of Earth. www.eoearth.org/view/article/155954.*

Why Is the Ocean Salty?

Seven friends were talking about the salty ocean. They wondered where most of the salt in the ocean comes from. They each had a different idea. This is what they said:

Tori: I think most of the salt comes from undersea vents and volcanoes.

Gus: I think most of the salt comes from rocks on the land.

Lila: I think most of the salt comes from rock and sediment on the ocean floor.

Arash: I think most of the salt comes from coral reefs.

Kaholo: I think the salt has always been there when the oceans were formed.

Daisy: I think most of the salt comes from the water cycle.

Malcom: I think most of the salt comes from melting icebergs and sea ice.

Who do you think has the best idea? _____ Explain your thinking.

_____

_____

_____

_____

Why Is the Ocean Salty?

Teacher Notes

Purpose
The purpose of this assessment probe is to elicit students' ideas about the ocean's salinity. The probe is designed to uncover students' ideas about the role of landforms and rivers in carrying salt into the ocean.

Type of Probe
Friendly talk

Related Concepts
Erosion, ocean, salinity, salt water, weathering

Explanation
The best answer is Gus's: "I think most of the salt comes from rocks on the land." Ocean salt comes primarily from rocks on the land. As rain falls, it contains some dissolved carbon dioxide from the air. This makes rainwater slightly acidic. As the rain falls on rock, it chemically weathers the rock and breaks it down into smaller particles and ions (particularly sodium and chloride ions). Those ions

in solution are carried away in runoff, small streams, and eventually rivers, which run into the ocean. The sodium and chloride ions make up more than 90% of all the dissolved ions in seawater.

Understanding why the sea is salty begins with knowing how water cycles among the ocean's physical states: liquid, vapor, and ice. As a liquid, in both freshwater and saltwater environments, water dissolves rocks and sediments and reacts with emissions from volcanoes and hydrothermal vents. This process creates a complex solution of mineral salts in the ocean basins, with most of the salts coming from rocks on land. Conversely, water vapor and ice are essentially salt free.

Water molecules cannot form crystals (ice) with salt ions present. When seawater freezes, the ice forms with pure water, leaving the salts behind to remain dissolved in the unfrozen sea water. Similarly, water vapor can form only as pure water without dissolved salts. Because the water in the ocean is salty, students commonly

think that the water cycle involving the ocean and the water cycle involving fresh water on land are different.

Administering the Probe

This probe is best used with grades 5–12. Emphasize to students that they should select the best answer to where *most* of the salt comes from because some answer choices, such as Tori's and Lila's, include sources of salt but do not account for most of the salt.

Related Core Ideas in *Benchmarks for Science Literacy* (AAAS 2009)

3–5 Processes That Shape the Earth

• Waves, wind, water, and ice shape and reshape the Earth's land surface by eroding rock and soil in some areas and depositing them in other areas, sometimes in seasonal layers.

6–8 The Earth

• Water evaporates from the surface of the Earth, rises and cools, condenses into rain or snow, and falls again to the surface. The water falling on land collects in rivers and lakes, soil, and porous layers of rock, and much of it flows back into the oceans.

6–8 Processes That Shape the Earth

• Rivers and glacial ice carry off soil and break down rock, eventually depositing the material in sediments or carrying it in solution to the sea.

Related Core Ideas in *A Framework for K–12 Science Education* (NRC 2012)

3–5 ESS2.A: Earth Materials and Systems

• Rainfall helps to shape the land and affects the types of living things found in a region. Water, ice, wind, living organisms, and gravity break rocks, soils, and sediments into smaller particles and move them around.

6–8 ESS2.C: The Roles of Water in Earth's Surface Processes

• Water continually cycles among land, ocean, and atmosphere via transpiration, evaporation, condensation and crystallization, and precipitation, as well as downhill flows on land.

9–12 ESS2.C: The Roles of Water in Earth's Surface Processes

• The abundance of liquid water on Earth's surface and its unique combination of physical and chemical properties are central to the planet's dynamics. These properties include water's exceptional capacity to absorb, store, and release large amounts of energy, transmit sunlight, expand upon freezing, dissolve and transport materials, and lower the viscosities and melting points of rocks.

Related Research

• A study by Freyberg (1985) revealed that many students think the Earth today is the same as it has always been.

• Feller (2007) identified several common misconceptions related to ocean water, including: (1) the ocean is salty because of deep sea vents, (2) all icebergs are made of saltwater, (3) the salty ocean is not linked to land's freshwater cycle.

Suggestions for Instruction and Assessment

• This probe can be combined with other probes such as "Beach Sand" (Keeley, Eberle, and Tugel 2007) and "Mountains and Beaches" (p. 111), which target the role of rivers in carrying rock and rock particles to the sea.

• Be aware that some students think of salt in a colloquial sense of the word—that is, the table salt we sprinkle on food. Although salt in ocean water does consist primarily of sodium chloride, help middle and high

school students develop the broader concept of salt, including sulfate, carbonate, calcium, and magnesium salts.

- Provide students with a map of the United States or larger land areas with river systems. Have them predict where the ocean would be the most salty (places where very little fresh water is entering the ocean system or evaporation is high), be the least salty (places where there is a large influx of fresh water to the ocean system from substantial river systems, large amounts of rainfall, or melting ice), and have mid-range salinity. The saltiest water is in the Red Sea and in the Persian Gulf, which have a salinity of about 4.1% because of very high evaporation rates and low freshwater influx). The least salty seas are in the polar regions, where both melting polar ice and a lot of rain dilute the salinity. Find a U.S. maps showing river systems at *http://www.mapsofworld.com/usa/usa-river-map.html*.

- Find a world map with river systems at *www.mapsofworld.com/thematic-maps/world-river-map.htmlhttp://www.mapsofworld.com/thematic-maps/world-river-map.html*.

- Have students investigate how the Great Salt Lake in Utah became salty. Compare similarities between the Great Salt Lake and the ocean in terms of how they became salty.

- Students can compare their predictions of salinity with this excellent video of the first global measurements of ocean salinity provided by the Aquarius satellite imagery from NASA. One feature to notice is a large patch of highly saline water across the North Atlantic. This area, the saltiest anywhere in the open ocean, is analogous to deserts on land, where little rainfall

and a lot of evaporation occur. Find the video at *www.nasa.gov/press-release/international-spacecraft-carrying-nasa-s-aquarius-instrument-ends-operations*.

- Students can investigate whether seawater can freeze under normal conditions at *http://aquarius.umaine.edu/cgi/ed_act.htm?id=12*. They can also evaporate saltwater to predict whether the salt evaporates with the water. (Safety notes: Have students wear sanitized goggles and aprons throughout the activity. Immediately wipe up any spilled water to avoid slips and falls. Be sure students wash their hands with soap and water after completing the activity.)

- Students can use the scientific and engineering practices of obtaining, evaluating, and communicating information related to this probe by accessing NOAA's National Ocean Service ocean facts at *http://oceanservice.noaa.gov/facts/whysalty.html*.

References

American Association for the Advancement of Science (AAAS). 2009. Benchmarks for science literacy online. *www.project2061.org/publications/bsl/online*.

Feller, R. 2007. 110 misconceptions about the ocean. *Oceanography* 20 (4). 170–173.

Freyberg, P. 1985. Implications across the curriculum. In *Learning in science*, ed. R. Osborne and P. Freyberg, 125–135. Auckland, New Zealand: Heinemann.

Keeley, P., F. Eberle, and J. Tugel. 2007. *Uncovering student ideas in science, vol. 2: 25 more formative assessment probes*. Arlington, VA: NSTA Press.

National Research Council (NRC). 2012. *A framework for K–12 science education: Practices, crosscutting concepts, and core ideas*. Washington, DC: National Academies Press.

Section 2
Water Cycle, Weather, and Climate

Concept Matrix... 46

**Related *Next Generation Science*
Standards Performance Expectations** ... 47

Related NSTA Resources 47

8 **Water Cycle Diagram**................................. 49

9 **Where Did the Water in the
Puddle Go?** 53

10 **Weather Predictors** 57

11 **In Which Direction Will the
Water Swirl?** 61

12 **Does the Ocean Influence Our
Weather or Climate?** 65

13 **Coldest Winter Ever!**................................. 69

14 **Are They Talking About Climate or
Weather?** 73

15 **What Are the Signs of
Global Warming?** 77

Concept Matrix
Probes #8–#15

CORE CONCEPTS ↓ / PROBES	#8 Water Cycle Diagram	#9 Where Did the Water in the Puddle Go?	#10 Weather Predictors	#11 In Which Direction Will the Water Swirl?	#12 Does the Ocean Influence Our Weather or Climate?	#13 Coldest Winter Ever!	#14 Are They Talking About Climate or Weather?	#15 What Are the Signs of Global Warming?
GRADE RANGE →	3–8	3–8	3–8	6–12	6–12	3–8	6–12	6–12
climate					X	X	X	X
climate change						X		X
Coriolis effect				X				
energy in the ocean system					X			
evaporation	X	X						
evidence								X
global climate					X			
global ocean conveyer belt					X			
global warming								X
global wind patterns				X				
humidity		X						
model	X							
ocean circulation				X	X			
ocean currents				X	X			
scale				X				
transpiration	X							
water cycle	X	X			X			
water vapor		X						
weather			X			X	X	X
weather forecasting			X					
winds				X				

Related *Next Generation Science Standards* Performance Expectations (NGSS Lead States 2013)

Earth's Systems

- Grade 5, 5-ESS2-1: Develop a model using an example to describe ways the geosphere, biosphere, hydrosphere, and/or atmosphere interact.
- Grades 6–8, MS-ESS2-4: Develop a model to describe the cycling of water through Earth's systems driven by energy from the sun and the force of gravity.
- Grades 9–12, HS-ESS2-2: Analyze geoscience data to make the claim that one change to Earth's surface can create feedbacks that cause changes to other Earth systems.
- Grades 9–12, HS-ESS2-4: Use a model to describe how variations in the flow of energy into and out of Earth's systems result in changes in climate.

Weather and Climate

- Kindergarten, K-ESS2-1: Use and share observations of local weather conditions to describe patterns over time.
- Grade 3, 3-ESS2-1: Represent data in tables and graphical displays to describe typical weather conditions expected during a particular season.
- Grades 6–8, MS-ESS2-5: Collect data to provide evidence for how the motions and complex interactions of air masses results in changes in weather conditions.
- Grades 6–8, MS-ESS3-5: Ask questions to clarify evidence of the factors that have caused the rise in global temperatures over the past century.
- Grades 9–12, HS-ESS2-4: Use a model to describe how variations in the flow of energy into and out of Earth's systems result in changes in climate.

- Grades 9–12, HS-ESS3-5: Analyze geoscience data and the results from global climate models to make an evidence-based forecast of the current rate of global or regional climate change and associated future impacts to Earth systems.

Reference

NGSS Lead States. 2013. *Next Generation Science Standards: For states, by states.* Washington, DC: National Academies Press. *www.nextgenscience.org/next-generation-science-standards*.

Related NSTA Resources
NSTA Press Books

Aretxabaleta, A., G. Brooks, and N. West. 2011. *Project earth science: Physical oceanography.* 2nd ed. Arlington, VA: NSTA Press.

Blake, R., J. A. Frederick, S. Haines, and S. C. Lee. 2010. *Inside-out: Environmental science in the classroom and the field, grades 3–8.* Arlington, VA: NSTA Press.

Constible, J., L. Sandro, and R. Lee. 2008. *Climate change from pole to pole: Biology investigations.* Arlington, VA: NSTA Press.

Environmental Literacy Council and National Science Teachers Association. 2007. *Global climate change: Resources for environmental literacy.* Arlington, VA: NSTA Press.

Kastens, K., and M. Turrin. 2010. *Earth science puzzles: Making meaning from data.* Arlington, VA: NSTA Press.

Konicek-Moran, R. 2013. *Everyday Earth and space science mysteries: Stories for inquiry-based science teaching.* Arlington, VA: NSTA Press.

Veal, W., and R. Cohen. 2011. *Project Earth science: Meteorology.* 2nd ed. Arlington, VA: NSTA Press.

NSTA Journal Articles

Beckrich, A. 2014. The green room: The latest on climate change. *The Science Teacher* 81 (5): 10.

Bowman, R. 2010. Idea bank: Climate inquiries. *The Science Teacher* 77 (2): 55–56.

Colaianne, B. 2015. Global warming. *The Science Teacher* 82 (1): 37–42.

German, S., and E. O'Day. 2009. Teaching: A reflective process. *Science Scope* 32 (9): 44–49.

Glen, M., and L. Smetana. 2010. Dress for the weather. *Science and Children* 47 (8): 32–35.

Golden, B., J. Grooms, V. Sampson, and R. Oliveri. 2012. Generating arguments about climate change. *Science Scope* 35 (7): 26–35.

Keeley, P. 2015. Wet jeans: Using familiar phenomena to uncover students' ideas. *Science and Children* 53 (2): 33–35.

Larson, B. 2010. Making the climate connection: Resources and learning progressions for teaching students about weather and climate change. *Science and Children* 33 (8): 61–65.

Pallant, A., H. Lee, and S. Pryputniewicz. 2012. Modeling Earth's climate. *The Science Teacher* 79 (7): 38–42.

Varelas, M., C. Pappas, A. Barry, and A. O'Neill. 2001. Examining language to capture scientific understandings: The case of the water cycle. *Science and Children* 38 (7): 26–29.

Vick, M. 2015. Tried and true: A water cycle of many paths. *Science Scope* 52 (5): 58–63.

NSTA Learning Center Resources
NSTA Webinars
List of NOAA Ocean and Climate Science–Related Webinars
http://learningcenter.nsta.org/products/web_seminar_archive_sponsor.aspx?page=NOAA
NGSS Core Ideas: Earth's Systems
http://learningcenter.nsta.org/products/symposia_seminars/NGSS/webseminar32.aspx

NSTA Science Objects
Earth, Moon, and Sun: General Characteristics of Earth
http://learningcenter.nsta.org/product_detail.aspx?id=10.2505/7/SCB-ESM.1.1
Ocean's Effect on Climate and Weather: Global Circulation Patterns
http://learningcenter.nsta.org/resource/?id=10.2505/7/SCB-OCW.1.3
Ocean's Effect on Weather and Climate: Changing Climate
http://learningcenter.nsta.org/resource/?id=10.2505/7/SCB-OCW.1.4
Ocean's Effect on Weather and Climate: Global Climate Patterns
http://learningcenter.nsta.org/resource/?id=10.2505/7/SCB-OCW.1.1
Ocean's Effect on Weather and Climate: Global Precipitation and Energy
http://learningcenter.nsta.org/resource/?id=10.2505/7/SCB-OCW.1.2

Water Cycle Diagram

Three students were working on a drawing of the water cycle for a class project. They each had different ideas about what needed to be in their drawing to show the water cycle. This is what they said:

Rema: There has to be an ocean in a water cycle diagram.

Van: There always has to be a body of water in a water cycle diagram. It doesn't have to be an ocean. It can be an ocean, lake, river, pond, or stream.

Lamar: A water cycle diagram doesn't have to have a body of water. We can draw it without an ocean, lake, river, pond, stream, or other body of water.

Which friend do you agree with the most? _____ Explain why you agree.

_____

_____

_____

_____

_____

Water Cycle Diagram

Teacher Notes

Purpose

The purpose of this assessment probe is to elicit students' ideas about the water cycle. The probe is designed to find out what students think a representation of the water cycle should include to show evaporation and transpiration.

Type of Probe

Friendly talk

Related Concepts

Evaporation, model, transpiration, water cycle

Explanation

The best answer is Lamar's: "A water cycle diagram doesn't have to have a body of water. We can draw it without an ocean, lake, river, pond, stream or other body of water." The water cycle involves evaporation of liquid water, condensation of water vapor, and precipitation (rain, sleet, hail, or snow). Water can evaporate from the ocean or other bodies of water as part

of the water cycle, but it can also evaporate as a result of the plant process of transpiration, animal respiration and waste, puddles, and soil. Textbook diagrams and posters of the water cycle often show evaporation occurring over a large body of water, such as an ocean or lake. They seldom show arrows indicating evaporation from living things, puddles, and the ground. As a result, some students believe that evaporation, as part of the water cycle, must include an ocean or other large body of water, and they fail to recognize other sources of evaporation, such as transpiration from plants.

Administering the Probe

This probe is best used with grades 3–8. This probe can be extended by using the formative assessment classroom technique of annotated drawing described on pages 4–5 (Keeley 2015). Ask students to draw and label a diagram of the water cycle to support their answer.

Related Core Ideas in *Benchmarks for Science Literacy* (AAAS 2009)

3–5 The Earth

- When liquid water disappears, it turns into a gas (vapor) in the air and can reappear as a liquid when cooled, or as a solid if cooled below the freezing point of water. Clouds and fog are made of tiny droplets or frozen crystals of water.

6–8 The Earth

- Water evaporates from the surface of the Earth, rises and cools, condenses into rain or snow, and falls again to the surface. The water falling on land collects in rivers and lakes, soil, and porous layers of rock, and much of it flows back into the oceans. The cycling of water in and out of the atmosphere is a significant aspect of the weather patterns on Earth.

Related Core Ideas in *A Framework for K–12 Science Education* (NRC 2012)

6–8 ESS2.C: The Roles of Water in Earth's Surface Processes

- Water continually cycles among land, ocean, and atmosphere via transpiration, evaporation, condensation and crystallization, and precipitation, as well as downhill flows on land.

Related Research

- A commonly held idea about the water cycle is that water evaporates only from an ocean or lake (Henriques 2000).
- In interviews with elementary students, some children thought seawater evaporates to form clouds (Bar 1989).
- In an analysis of students' drawings of the water cycle, researchers found that student drawings usually showed evaporation from

a large body of water, such as the ocean (Osman 2009).

- A study of junior high students' perceptions of the water cycle revealed that most students were aware of the atmospheric part of the water cycle but ignored its groundwater part. Moreover, students who included part of the underground system in the water cycle perceived the underground water as static, sub-surface lakes (Ben-zvi-Assarf and Orion 2005).

Suggestions for Instruction and Assessment

- Have students create a representation of the water cycle as it would occur in an area that does not have an ocean, lake, or body of water.
- Have students tie a plastic bag around a leafy branch or shrub that is in the sun and leave it for a day to observe transpiration. Have students observe the water inside the bag and have students explain where the water came from and where it goes when it leaves the plant.
- Have students conduct an online image search for representations of the water cycle. Have them look for images that include evaporation sources other than just large bodies of water.
- Explicitly point out to students the limitations of models such as drawings. Explain that models cannot always portray all aspects of the real thing.

References

American Association for the Advancement of Science (AAAS). 2009. Benchmarks for science literacy online. *www.project2061.org/publications/bsl/online*.

Bar, V. 1989. Children's views about the water cycle. *Science Education* 73 (4): 481–500.

Ben-zvi-Assarf, O., and N. Orion. 2005. A study of junior high students' perceptions of the

8

water cycle. *Journal of Geoscience Education* 53 (4): 366–373.

Henriques, L. 2000. Children's misconceptions about weather: A review of the literature. Paper presented at the annual meeting of the National Association of Research in Science Teaching, New Orleans, LA.

Keeley, P. 2015. *Science formative assessment: 75 practical strategies for linking assessment,* *instruction, and learning.* 2nd ed. Thousand Oaks, CA: Corwin Press.

National Research Council (NRC). 2012. *A framework for K–12 science education: Practices, crosscutting concepts, and core ideas.* Washington, DC: National Academies Press.

Osman, C. 2009. Science students' misconceptions of the water cycle according to their drawings. *Journal of Applied Sciences* 9 (5): 865–873.

Where Did the Water in the Puddle Go?

Six friends were walking to the park on a sunny day. They noticed a big puddle on their way to the park. When they came back two hours later, the puddle was gone. They each had different ideas about where the water in the puddle went. This is what they said:

Bandar: The water went right up to the clouds.

Regina: The sun dried the water up and it no longer exists.

Dylan: All of the water soaked into the ground.

Axel: Much of the water is in the air around us.

Minato: The water went into a stream, river, pond, lake, or ocean.

Clint: The water changed into fog.

Which friend do you think has the best idea? _____ Explain your thinking.

_____

_____

_____

_____

Where Did the Water in the Puddle Go?

Teacher Notes

Purpose

The purpose of this assessment probe is to elicit students' ideas about evaporation. The probe is designed to find out if students recognize that water goes into the air around us in the form of water vapor.

Type of Probe

Friendly talk, familiar phenomenon

Related Concepts

Evaporation, humidity, water cycle, water vapor

Explanation

The best answer is Axel's: "Much of the water is in the air around us." Although some of it will soak into the ground, much of it evaporates. As the water evaporates, it becomes a gas (water vapor), which we cannot see. The water never ceases to exist, as explained by the law of conservation of matter. Contrary to what is often represented in water cycle diagrams with big, long arrows that point from water to a cloud, water does not immediately go up to the clouds. It rises and exists in the air around us as an invisible gas. Eventually, some water molecules will rise high in the atmosphere, condense, and form clouds, but most of the water exists in the air around us as part of a cycle of evaporation and condensation. Humid weather provides evidence of water vapor in the air, including explanatory phenomena of condensation, such as wet dew on the grass in the morning or water on the outside of a cold beverage glass.

Administering the Probe

This probe is best used with grades 3–8. Tell the students that the puddle was on a paved area of the park. If possible, you may even create a puddle on a sidewalk or pavement in the schoolyard so that students can observe the puddle and then come back later to observe that the water is gone. This probe can be extended by having students draw a picture to explain their thinking.

Related Core Ideas in *Benchmarks for Science Literacy* (AAAS 2009)

3–5 The Earth

- When liquid water disappears, it turns into a gas (vapor) in the air and can reappear as a liquid when cooled, or as a solid if cooled below the freezing point of water. Clouds and fog are made of tiny droplets or frozen crystals of water.

6–8 The Earth

- Water evaporates from the surface of the Earth, rises and cools, condenses into rain or snow, and falls again to the surface. The water falling on land collects in rivers and lakes, soil, and porous layers of rock, and much of it flows back into the oceans. The cycling of water in and out of the atmosphere is a significant aspect of the weather patterns on Earth.

Related Core Ideas in *A Framework for K–12 Science Education* (NRC 2012)

3–5 PS1.A: Structure and Properties of Matter

- Matter of any type can be subdivided into particles that are too small to see, but even then the matter still exists and can be detected by other means. A model showing that gases are made from matter particles that are too small to see and are moving freely around in space can explain many observations, including the inflation and shape of a balloon and the effects of air on larger particles or objects.

6–8 ESS2.C: The Roles of Water in Earth's Surface Processes

- Water continually cycles among land, ocean, and atmosphere via transpiration, evaporation, condensation and crystallization, and precipitation, as well as downhill flows on land.

Related Research

- Research has shown that students seem to go through a series of stages before they fully understand evaporation as a process that converts water to an invisible form. At first, they may seem to think that when water evaporates, it simply ceases to exist. In the next stage, they may think it changes location but changes into some other form we can perceive, such as fog, steam, or droplets. Fifth grade is about the time that students can accept air as the location of evaporating water if they have had special instruction that targets this idea (AAAS 2009).

- Students' scientific concept of evaporation appears to be dependent on three notions: (1) conservation of matter, (2) the idea of atoms or molecules, and (3) the idea that air contains particles we cannot see. At ages 12–14, students are apt to link those three ideas (Driver et al. 1994).

- A study by Barr and Travis (1991) found that students' understanding of evaporation in a boiling context may precede their understanding of evaporation of water from surfaces. In the sample in that study, 70% of six- to eight-year-old students understood that there was vapor coming out of water when it boiled. They understood that (1) the water was going somewhere because the amount decreased from the container and (2) the vapor was made of water. However, the same children thought that when a solid, wet object dried, the water simply disappeared or it went into the object (Driver et al. 1994).

Suggestions for Instruction and Assessment

- Use the probe, "Wet Jeans" as a post-assessment to check on understanding after students have had the opportunity to develop their ideas (Keeley, Eberle, and Farrin 2005).

- Have students generate a list of phenomena, such as the wet dew on the grass in the morning, that provide explanatory evidence that water is in the air around us in a form we cannot see.

- Elementary students should have multiple experiences in observing how water "disappears" from various surfaces and open containers. Older elementary students should be challenged to think about where the water goes immediately after it disappears and where it may eventually end up.

- Be careful when using language such as "the water disappeared." Use an analogy such as a student leaving the room. We can say the student "disappeared" because we can no longer see the student. However, the student still exists somewhere. It is important to be careful how we use words such as *disappear*. Students may develop incorrect notions of conservation of matter in the context of the water cycle if teachers are not careful with the words they use to describe phenomena.

- Teach students what happens to water before introducing terminology such as evaporation and water vapor. Many students will use the word evaporation without understanding where water actually goes. Students need inquiry-based experiences to discover for themselves that water is in the air around them before using the terminology. The careful wording in *Benchmarks for Science Literacy* (AAAS 1993, 2009) affirms this notion of developing some ideas before using terminology. Once students have experienced the concept, the term can be introduced with meaning.

- A-B-C-C-B-V is a strategy that supports conceptual understanding: engage students in an activity with phenomena (figuring out where the water went) before introducing a concept (water in the air in a form we cannot see). Connect the activity to the concept before introducing the vocabulary (evaporation and water vapor). Once they have the concept, the vocabulary can be attached to it.

- Be aware that some students may not identify the puddle as being part of the water cycle. This failure to place puddles in the water cycle may result from the student seeing representations that usually show water evaporating from a large body of water. Combine this probe with "Water Cycle Diagram" to see if student thinking is limited to evaporation from large bodies of water only.

- Explicitly point out water cycle diagrams that show large arrows labeled evaporation pointing up to a cloud. Explain that the diagram is showing the long-term cyclic process but is not intended to imply that water immediately goes up to a cloud.

References

American Association for the Advancement of Science (AAAS). 1993. *Benchmarks for science literacy.* New York: Oxford University Press.

American Association for the Advancement of Science (AAAS). 2009. Benchmarks for science literacy online. *www.project2061.org/publications/bsl/online.*

Barr, V., and A. Travis. 1991. Children's views concerning phase changes. *Journal of Research in Science Teaching* 28 (4): 363–382.

Driver, R., A. Squires, P. Rushworth, and V. Wood-Robinson. 1994. *Making sense of secondary science: Research into children's ideas.* London: RoutledgeFalmer.

Keeley, P., F. Eberle, and L. Farrin. 2005. *Uncovering student ideas in science, vol. 1: 25 formative assessment probes.* Arlington, VA: NSTA Press.

National Research Council (NRC). 2012. *A framework for K–12 science education: Practices, crosscutting concepts, and core ideas.* Washington, DC: National Academies Press.

 10

Weather Predictors

Five friends went camping in Northern Michigan in July. They wondered what the weather would be like that winter. They each had different ideas about long-term weather predicting. This is what they said:

Yong: I think we can predict the winter weather by how hot it is right now in July.

Saul: I don't think we can predict what this year's winter weather will be like based on just our summer weather.

Mark: I think there are natural signs from animals that can be used to predict the winter weather, such as the thickness of fur and color bands and the number of wooly caterpillars.

Tad: My parents have a book that predicts what the winter weather will be like. It's called *The Farmer's Almanac*.

Jorge: I think the groundhog's shadow is a good predictor of winter weather.

Who do you think has the best idea? _____ Explain your thinking.

_____

_____

_____

_____

Weather Predictors

Teacher Notes

Purpose

The purpose of this assessment probe is to elicit students' ideas about weather predictions. The probe is designed to find out if students recognize the limitations of long-term weather forecasting.

Type of Probe

Friendly talk

Related Concepts

Weather, weather forecasting

Explanation

The best answer is Saul's: "I don't think we can predict what this year's winter weather will be like based on just our summer weather." Long-term weather predictions can be made, but they involve several factors besides what the summer weather is like. It is difficult to make accurate, long-term weather predictions several months in advance with limited data. Predictions and weather forecasts consider a variety of factors in

the atmosphere—not animal behavior, thickness of fur, a ground hog's shadow, or how hot the preceding summer is. Some animals (e.g., wooly caterpillars) experience changes in numbers and coloration that seem to predict the severity of an upcoming winter. Although those patterns have sometimes been noticed to correlate with harsh winters, entomologists generally agree that woolly caterpillars are not accurate predictors of winter weather. Many variables may contribute to changes in the wooly caterpillar's numbers and coloration, such as food availability, temperature or moisture during larval development, age, and even differences in species.

The Farmer's Almanac has an interesting history of predicting long-term weather that combines sunspot activity, climatology, and meteorology. The resource uses 30-year statistical averages and claims accuracy up to 80%. Some of their predictions hold true, whereas others are very wrong. Meteorologists generally do not consider *The Farmer's Almanac* a scientific predictor of long-range weather forecasts.

Meteorologists do make seasonal weather predictions based on factors such as the El Niño and La Niña cycle and the Arctic Oscillation. The issue is the amount of uncertainty inherent in the prediction; the further ahead in time the prediction is made, the higher the uncertainty. The important point for students to know is that scientists look at various factors and patterns to make short- and long-term predictions about weather with various degrees of accuracy.

Administering the Probe

This probe is best used with grades 3–8. Make sure students know that the boys are wondering about what the specific winter weather will be like months before the coming winter.

Related Core Ideas in *Benchmarks for Science Literacy* (AAAS 2009)

K–2 The Earth
- The temperature and amount of rain (or snow) tend to be high, low, or medium in the same months every year.

3–5 The Earth
- The weather is always changing and can be described by measurable quantities such as temperature, wind direction and speed, and precipitation. Large masses of air with certain properties move across the surface of the Earth. The movement and interaction of these air masses is used to forecast the weather.

6–8 The Earth
- The temperature of a place on the Earth's surface tends to rise and fall in a somewhat predictable pattern every day and over the course of a year. The pattern of temperature changes observed in a place tends to vary depending on how far north or south of the equator the place is, how near to oceans it is, and how high above sea level it is.

Related Core Ideas in *A Framework for K–12 Science Education* (NRC 2012)

K–2 ESS2.D: Weather and Climate
- Weather is the combination of sunlight, wind, snow or rain, and temperature in a particular region at a particular time. People measure these conditions to describe and record the weather and to notice patterns over time.

3–5 ESS2.D: Weather and Climate
- Scientists record patterns of the weather across different times and areas so that they can make predictions about what kind of weather might happen next.

6–8 ESS2.D: Weather and Climate
- Weather and climate are influenced by interactions involving sunlight, the ocean, the atmosphere, ice, landforms, and living things. These interactions vary with latitude, altitude, and local and regional geography, all of which can affect oceanic and atmospheric flow patterns.
- Because these patterns are so complex, weather can only be predicted probabilistically.

Related Research
- Some students may think that winter weather can be predicted by studying the thickness of the fur on some animals (Phillips 1991).
- A study of preservice teachers' ideas revealed that some believed that very cold winters can be predicted by seeing how hot it was the previous summer (Schoon 1995).

Suggestions for Instruction and Assessment
- Follow up by having students develop rebuttals to explain why each of the distracters is not an accurate way to predict long-term weather.

- Help students understand the difference between weather in the short term versus climate in the long term. Climate allows us to predict what the weather will generally be like months in advance, based on long-term patterns. However, climate cannot predict with 100% accuracy what the specific weather conditions will be like months in advance.

- Have students interview a meteorologist to find out how far in advance she or he can generally make accurate predictions about weather.

- Have students research and present weather forecasting folklore and proverbs and contrast them with what is considered scientific evidence.

- Have students research the history of and how *The Farmer's Almanac* predictions are made. Students can debate the scientific accuracy of the almanac's prediction method as well as whether the secrecy of its "prediction formula" is in keeping with the nature of science. Tie in the mathematics of probability to the predictions made by *The Farmer's Almanac*.

- Older students can study some of the ways scientists use data and patterns to research and predict seasonal weather and why those predictions are important to certain groups. They can begin their research at the National Science Foundation's page on predicting seasonal weather at *www.nsf.gov/news/special_reports/autumnwinter/intro.jsp*.

References

American Association for the Advancement of Science (AAAS). 2009. Benchmarks for science literacy online. *www.project2061.org/publications/bsl/online*.

National Research Council (NRC). 2012. *A framework for K–12 science education: Practices, crosscutting concepts, and core ideas*. Washington, DC: National Academies Press.

Philips, W. 1991. Earth science misconceptions. *The Science Teacher* 58 (2): 21–23.

Schoon, K. 1995. The origin and extent of alternative conceptions in the Earth and space sciences: A survey of pre-service elementary teachers. *Journal of Elementary Science Education* 7 (2): 27–46.

In Which Direction Will the Water Swirl?

Three friends were wondering if the direction water swirled down a drain depended on location. Avery lives in the Northern Hemisphere. Tara lives on the Equator. Magdelena lives in the Southern hemisphere. They predicted what would happen if Avery and Magdelena filled their bathtubs with water at the same time, pulled the plug, and observed the direction of water as it went down the drain. This is what they said:

Avery: I think the water in my bathtub will swirl down the drain in the opposite direction of the water in Magdelena's bathtub.

Magdelena: I think the location doesn't make a difference in the direction water swirls down a drain.

Tara: I think water always swirls down a drain in the same direction.

Who do you agree with the most? _____ Explain why you agree.

_____

_____

_____

_____

In Which Direction Will the Water Swirl?

Teacher Notes

Purpose

The purpose of this assessment probe is to elicit students' ideas about the Coriolis effect. The probe is designed to find out whether students recognize how scale and Earth's rotation affect the behavior of fluids.

Type of Probe

Friendly talk

Related Concepts

Coriolis effect, global wind patterns, ocean circulation, ocean currents, scale, winds

Explanation

The best answer is Magdelena's: "I think the location doesn't make a difference in the direction water swirls down a drain." A common misconception is that water swirls down a drain in the opposite direction in the opposite hemisphere. Some people attribute this to a misunderstanding of the Coriolis effect. The

Coriolis effect describes how rotating objects are deflected. It helps explain the movement of large fluid masses. Since Earth rotates on its axis, circulating air is deflected to the right (clockwise spiral) in the Northern hemisphere and the left (counterclockwise spiral) in the Southern hemisphere. This deflection affects the direction of large bodies of fluids (liquid and gas) such as global winds and ocean currents. The slow rotation of Earth means the Coriolis effect is not strong enough, despite a popular myth, to be seen in small movements and volumes of water, such as the draining of water in a bathtub or sink or the flushing of water down a toilet.

Administering the Probe

This probe is best used with grades 6–12. It should be used when students are learning about global wind patterns or ocean circulation. Consider pointing out the geographic location of each person on a map before students respond to the probe.

Related Core Ideas in *Benchmarks for Science Literacy* (AAAS 2009)

9–12 The Earth

- Transfer of thermal energy between the atmosphere and the land or oceans produces temperature gradients in the atmosphere and the oceans. Regions at different temperatures rise or sink or mix, resulting in winds and ocean currents. These winds and ocean currents, which are also affected by the Earth's rotation and the shape of the land, carry thermal energy from warm to cool areas.

9–12 Scale

- Because different properties are not affected to the same degree by changes in size, large changes in size typically change the way that things work in physical, biological, or social systems.

Related Core Ideas in *A Framework for K–12 Science Education* (NRC 2012)

6–8 ESS2.A: Earth's Materials and Systems

- The planet's systems interact over scales that range from microscopic to global in size, and they operate over fractions of a second to billions of years.

6–8 ESS2.C: The Roles of Water in Earth's Surface Processes

- The complex patterns of the changes and the movement of water in the atmosphere, determined by winds, landforms, and ocean temperatures and currents, are major determinants of local weather patterns.

Related Research

- There is no formal research on this misconception, but it is a widely held myth of both students and adults. The author (Page Keeley) actually witnessed this very

common belief firsthand during a trip to Ecuador, where she observed tourists standing at the demarcation of the Equator line interacting with locals who devised a clever tourist trap. Holding basins of water, they demonstrated (for a donation) how the water swirled in one direction on one side of the Equator and then stepping over the line, they demonstrated how it swirled in the opposite direction. Tourists were delighted to see their observation confirm their beliefs and gladly gave the local "entertainers" a donation. I carefully observed how adept the locals were at cleverly masking their manipulation of the basins!

Suggestions for Instruction and Assessment

- The Coriolis effect is not a big idea or serious misconception that interferes with how students connect science concepts. However, because this misconception is so pervasive, it is interesting to use it as an initial elicitation and discussion before students learn about the movement patterns of winds and ocean currents.

- This probe is an opportune time to address the crosscutting concept of scale, particularly how things in the natural world behave at different scales.

- Have students research the myth of the Coriolis effect and describe at what scale it does and does not have an effect on systems.

References

American Association for the Advancement of Science (AAAS). 2009. Benchmarks for science literacy online. *www.project2061.org/publications/bsl/online*.

National Research Council (NRC). 2012. *A framework for K–12 science education: Practices, crosscutting concepts, and core ideas.* Washington, DC: National Academies Press.

Does the Ocean Influence Our Weather or Climate?

Four friends were talking about the ocean. They each had different ideas about how the ocean influences conditions on Earth. This is what they said:

Justin: The ocean has a major influence on climate, but I don't think it has much of an effect on the weather.

Bruce: The ocean has a major influence on weather, but I don't think it has much of an effect on climate.

Maya: I disagree with both of you. The ocean has a major influence on both the weather and climate.

Para: I disagree with all of you. The ocean does not have a major influence on weather or climate. It's just part of the water cycle.

Who do you agree with the most? _____ Explain why you agree.

_____

_____

_____

_____

Does the Ocean Influence Our Weather or Climate?

Teacher Notes

Purpose

The purpose of this assessment probe is to elicit students' ideas about the connection between the oceans and weather and climate. The probe is designed to reveal how students think the ocean influences atmospheric conditions.

Type of Probe

Friendly talk

Related Concepts

Climate, energy in the ocean system, global climate, global ocean conveyer belt, ocean circulation, ocean currents, water cycle, weather

Explanation

Maya has the best idea: "I disagree with both of you. The ocean has a major influence on both the weather and climate." Of all the influences on Earth's weather and weather patterns, the ocean plays one of the most significant roles in determining the weather and climate. Although the Sun and gravity may

be the driving forces behind the movement of water, the ocean and its currents create weather patterns over time, as well as localized weather events. The ocean also plays a fundamental role in shaping the climate zones we see on land. Even areas hundreds of miles away from any coastline are still largely influenced by the global ocean system.

Ocean currents act much like a conveyer belt, transporting warm water and precipitation from the equator to the poles and transporting cold water from the poles back to the tropics. Thus, currents regulate global climate, helping to counteract the uneven distribution of solar radiation reaching Earth's surface. Without currents, regional temperatures would be more extreme—super hot at the equator and frigid toward the poles—and much less of Earth's land would be habitable.

The world's ocean is crucial to heating the planet. Although land areas and the atmosphere absorb some sunlight, the ocean absorbs the majority of the Sun's radiation. Particularly

in the tropical waters around the equator, the ocean acts as a massive heat sink.

The ocean doesn't just store solar radiation, it also helps to distribute heat around the globe. When molecules of liquid water gain enough thermal energy, they escape into the air in the form of water vapor. Ocean water is constantly evaporating, thereby increasing the temperature and humidity of the surrounding air to form rain and storms that are then carried by winds, often vast distances. In fact, almost all rain that falls on land starts off in the ocean. The tropics are particularly rainy because of the amount of heat absorption, and thus the rate of ocean evaporation, is highest in this area.

Administering the Probe

This probe is best used with grades 6–12. Make sure students understand that influence means having an effect on temperatures, wind patterns, and precipitation.

Related Core Ideas in *Benchmarks for Science Literacy* (AAAS 2009)

6–8 The Earth

- Water evaporates from the surface of the Earth, rises and cools, condenses into rain or snow, and falls again to the surface. The water falling on land collects in rivers and lakes, soil, and porous layers of rock, and much of it flows back into the oceans. The cycling of water in and out of the atmosphere is a significant aspect of the weather patterns on Earth.
- The Earth has a variety of climates, defined by average temperature, precipitation, humidity, air pressure, and wind, over time in a particular place.
- Thermal energy carried by ocean currents has a strong influence on climates around the world. Areas near oceans tend to have

more moderate temperatures than they would if they were farther inland but at the same latitude because water in the oceans can hold a large amount of thermal energy.

Related Core Ideas in *A Framework for K–12 Science Education* (NRC 2012)

3–5 ESS2.D: Weather and Climate

- Climate describes a range of an area's typical weather conditions and the extent to which those conditions vary over years.

6–8 ESS2.D: Weather and Climate

- Weather and climate are influenced by interactions involving sunlight, the ocean, the atmosphere, ice, landforms, and living things. These interactions vary with latitude, altitude, and local and regional geography, all of which can affect oceanic and atmospheric flow patterns.
- The ocean exerts a major influence on weather and climate by absorbing energy from the Sun, releasing it over time, and globally redistributing it through ocean currents.

9–12 ESS2.D: Weather and Climate

- The foundation for Earth's global climate systems is the electromagnetic radiation from the Sun, as well as its reflection, absorption, storage, and redistribution among the atmosphere, ocean, and land systems, and this energy's re-radiation into space.

Related Research

- Feller (2007) identified several weather- and climate-related misconceptions people have about the ocean, including the idea that it doesn't rain as much over the ocean as it does on land and that winds and sunlight control the climate much more than do ocean currents or sea-surface temperature.

Suggestions for Instruction and Assessment

- Students can research the conditions that create the El Niño and La Niña weather patterns that affect the western United States. If a strong El Niño and La Niña pattern is predicted, students can collect weather data to support or disprove the prediction.
- Visit the NOAA's Ocean Service Education (*http://oceanservice.noaa.gov/education/pd/oceans_weather_climate/welcome.html*), which provides resources for teachers and students that include videos, graphs, and interactive modules on ocean currents, hurricane paths, and so on.
- Examine the Ocean Literacy Framework at *http://oceanliteracy.wp2.coexploration.org*. The Framework includes the Seven Principles of Ocean Literacy, including Principle #3: "The ocean is a major influence on weather and climate." The Framework also includes a scope and sequence and graphical conceptual flow maps to understand how the ocean affects weather and climate.

References

American Association for the Advancement of Science (AAAS). 2009. Benchmarks for science literacy online. *www.project2061.org/publications/bsl/online*.

Feller, R. 2007. 110 misconceptions about the ocean. *Oceanography* 20 (4): 170–173.

National Research Council (NRC). 2012. *A framework for K–12 science education: Practices, crosscutting concepts, and core ideas*. Washington, DC: National Academies Press.

Coldest Winter Ever!

Mr. and Mrs. Martin were watching the news one evening in January. The news anchor announced that Florida was experiencing one of the coldest winters ever recorded in that state. The temperature even dropped to a low of 32°F in Miami. It snowed in some parts of Florida, and crops were in danger of freezing. Mr. and Mrs. Martin looked at each other and said the following:

Mr. Martin: Wow! That is a big change in our weather!

Mrs. Martin: Wow! That is a big change in our climate!

Which person do you agree with the most? _____ Explain why you agree.

_____

_____

_____

_____

_____

Coldest Winter Ever!

Teacher Notes

Purpose

The purpose of this assessment probe is to elicit students' ideas about weather versus climate. The probe is designed to find out if students distinguish between a one-time weather-related change and a change in climate.

Type of Probe

Opposing views

Related Concepts

Climate, climate change, weather

Explanation

The best answer is Mr. Martin's: "Wow! That is a big change in our weather!" Weather and climate are related, but they are different things. The main difference is time. Weather varies from day to day and seasonally. *Weather* is the short-term conditions of the atmosphere, such as temperature, humidity, precipitation, cloudiness, and wind that can vary from hour to hour, day to day, week to week, and season to season.

The cold spell in Florida is an example of an extreme one-time weather event that occurred that year, and some people use weather events such as this to falsely dispel the idea of climate change and global warming. It is an unusual change in the weather for that time of year and location and does not occur as a pattern year after year. *Climate* describes a long-term pattern of weather over many years, from decades to millions of years, for a particular region. A drastic change that happens during one winter does not describe climate. However, if this cold winter in Florida continued as a pattern for many years, it could be described as climate. Climate scientists use a minimum of 30 years to define a region's climate.

Administering the Probe

This probe is best used with grades 3–8. This probe can be combined with the formative assessment classroom technique of lines of agreement (see pp. 6–9). Make sure students know what the average weather is like in Miami, Florida.

Related Core Ideas in *Benchmarks for Science Literacy* (AAAS 2009)

K–2 The Earth

- The temperature and amount of rain (or snow) tend to be high, low, or medium in the same months every year.

3–5 The Earth

- The weather is always changing and can be described by measurable quantities such as temperature, wind direction and speed, and precipitation. Large masses of air with certain properties move across the surface of the Earth. The movement and interaction of these air masses is used to forecast the weather.

6–8 The Earth

- The temperature of a place on the Earth's surface tends to rise and fall in a somewhat predictable pattern every day and over the course of a year. The pattern of temperature changes observed in a place tends to vary depending on how far north or south of the equator the place is, how near to oceans it is, and how high above sea level it is.
- The Earth has a variety of climates, defined by average temperature, precipitation, humidity, air pressure, and wind, over time in a particular place.

Related Core Ideas in *A Framework for K–12 Science Education* (NRC 2012)

K–2 ESS2.D: Weather and Climate

- Weather is the combination of sunlight, wind, snow or rain, and temperature in a particular region at a particular time. People measure these conditions to describe and record the weather and to notice patterns over time.

3–5 ESS2.D: Weather and Climate

- Climate describes a range of an area's typical weather conditions and the extent to which those conditions vary over years.

Related Research

- In a study conducted by the Yale Project on Climate Change Communication, 80% of middle and high school students thought it was true, or mostly true, that climate changes from year to year (Leiserowitz, Smith, and Marlon 2011).

Suggestions for Instruction and Assessment

- Students can collect weather data over several days or weeks, graph temperature data, and compare the temperature data collected with long-term climate averages from where they live.
- Students can explore the University Corporation for Atmospheric Research's "Kids Crossing" website, which helps children understand the difference between weather and climate (available at *https://eo.ucar.edu/ kids/green/what1.htm*).
- Have students choose a time of the year and an area of the world they would like to travel to. Have students describe what kinds of clothes would be suitable to pack for that time of year. Discuss whether an understanding of weather or climate helped them decide what kinds of clothes to bring.
- *Climate Literacy: The Essential Principles of Climate Science* is a guide for educators who teach climate science. This guide can be accessed at *www.globalchange.gov/ browse/educators*.
- Have students select a city and analyze and compare three to five years of weather data for the same two weeks of a month. Have students describe patterns they see from year to year over the same two-week

period. The Weather Underground website (*www.wunderground.com/history*) archives historical weather data that students can search by city.

- The probe can be extended to middle school and high school by engaging students in a discussion of whether the cold spell can be used as evidence that disproves the claim that Earth is warming. Listen for whether students reason that (1) this is a one-time weather event, not a climate change and (2) this is a localized event, not an event occurring globally over time.

References

American Association for the Advancement of Science (AAAS). 2009. Benchmarks for science literacy online. *www.project2061.org/publications/bsl/online.*

Leiserowitz, A., Smith, N., and Marlon, J. R. 2011. American teens' knowledge of climate change. Yale Project on Climate Change Communication, Yale University. *http://environment.yale.edu/climate-communication/article/american-teens-knowledge-of-climate-change.*

National Research Council (NRC). 2012. *A framework for K–12 science education: Practices, crosscutting concepts, and core ideas.* Washington, DC: National Academies Press.

Are They Talking About Climate or Weather?

People like to talk about conditions that affect their lives. Climate and weather are two of those conditions. Put a C next to all the things people say that have to do with climate. Put a W next to all the things people say that have to do with weather.

_____ **A.** "What shall I wear today?"

_____ **B.** "What equipment do I need for the camping trip next summer?"

_____ **C.** "It has rained on my birthday for the past three years."

_____ **D.** "We just got three feet of snow in March. So much for global warming!"

_____ **E.** "I wonder when I should plant the tomato seeds in my garden?"

_____ **F.** "Our state has experienced the worst drought since records began more than 120 years ago."

_____ **G.** "News flash! The drought in California has ended with the first significant storm this year dumping more than 10 inches of rain in many locations and filling most of the reservoirs to pre-drought conditions."

_____ **H.** "We can't get low-cost flood insurance for our house anymore. The insurance company says this area is too great a risk."

Explain your thinking. What rule or reasoning did you use to decide if a statement is related to climate or weather?

_____

_____

_____

_____

Related Core Ideas in *Benchmarks for Science Literacy* (AAAS 2009)

3–5 The Earth

- The weather is always changing and can be described by measurable quantities such as temperature, wind direction and speed, and precipitation.

6–8 The Earth

- The Earth has a variety of climates, defined by average temperature, precipitation, humidity, air pressure, and wind, over time in a particular place.

Related Core Ideas in *A Framework for K-12 Science Education* (NRC 2012)

K–2 ESS2.D: Weather and Climate

- Weather is the combination of sunlight, wind, snow or rain, and temperature in a particular region at a particular time. People measure these conditions to describe and record the weather and to notice patterns over time.

3–5 ESS2.D: Weather and Climate

- Climate describes a range of an area's typical weather conditions and the extent to which those conditions vary over years.

6–8 ESS2.D: Weather and Climate

- Weather and climate are influenced by interactions involving sunlight, the ocean, the atmosphere, ice, landforms, and living things. These interactions vary with latitude, altitude, and local and regional geography, all of which can affect oceanic and atmospheric flow patterns.

Related Research

- In a study conducted by the Yale Project on Climate Change Communication, 80% of middle and high school students thought it was true, or mostly true, that climate

changes from year to year (Leiserowitz, Smith, and Marlon 2011).

Suggestions for Instruction and Assessment

- This probe can be used with the formative assessment card sort strategy (Keeley 2008, 2015). Print each of the statements on cards and have students work in small groups to sort them into statements related to climate and statements related to weather. During the card sort, have students justify their reasoning. Listen carefully for evidence of their distinguishing between climate-related ideas and weather-related ideas. After instructional experiences have been provided to help students distinguish between weather and climate, provide them with an opportunity to re-sort the cards and explain how their thinking changed.
- Students can collect weather data over several days or weeks, graph temperature data, and compare the temperature data collected with long-term climate averages from where they live. Understanding the difference between weather and climate and interpreting local weather data are important first steps to understanding larger-scale global climate changes. Data are available at *http://cleanet.org/resources/42787.html.*
- A video on PBS Learning Media's website features University of Wisconsin at Madison researcher John Magnuson. He explains the difference between weather and climate using data on ice cover from Lake Mendota in Madison, Wisconsin. Analysis of the data indicates a long-term trend that can be connected to climate change. The video is available at *www.pbslearningmedia. org/resource/ecb10.sci.ess.watcyc.weather/ the-difference-between-weather-and-climate.*
- NASA and NOAA have websites that explain the difference between weather and climate. You can access the NASA

site at *www.nasa.gov/mission_pages/noaa-n/ climate/climate_weather.html* and the NOAA site at *http://oceanservice.noaa.gov/ facts/weather_climate.html.*

- Climate Literacy: The Essential Principles of Climate Science is a guide for educators who teach climate science. This guide can be accessed at *www.globalchange.gov/ browse/educators.*

- For high school students, consider adding two statements that are inconclusive as to whether they are about climate or weather. For example, "Back when I was a boy in 1937, we had a bad bark beetle problem that killed most of the trees, I think. Now they've returned this year in even bigger numbers, killing even more trees." and "Do you ever remember it being this cold before?" The comparison of this particular bark beetle outbreak to climate change does not indicate a connection. The outbreak could be due to a number of reasons, such as a particularly warm or dry year. Insect outbreaks respond to subtle changes in weather, but they can also respond to changes over time, which may or may not be due to changes in the climate. Memory of a particularly cold spell can be inaccurate. For instance, news reports during the winter of 2013–2014

said that it was the coldest winter on the East Coast in recorded history. But it had actually been that cold just 20 years before. It is always best to use reliable data when drawing conclusions.

References

American Association for the Advancement of Science (AAAS). 2009. Benchmarks for science literacy online. *www.project2061.org/ publications/bsl/online.*

Keeley, P. 2008. *Science formative assessment: 75 practical strategies for linking assessment, instruction, and learning.* Thousand Oaks, CA: Corwin Press and Arlington, VA: NSTA Press.

Keeley, P. 2015. *Science formative assessment: 75 practical strategies for linking assessment, instruction, and learning.* 2nd ed. Thousand Oaks, CA: Corwin Press.

Leiserowitz, A., Smith, N., and Marlon, J. R. 2011. American teens' knowledge of climate change. Yale Project on Climate Change Communication, Yale University. *http://environment.yale.edu/climate-communication/article/ american-teens-knowledge-of-climate-change.*

National Research Council (NRC). 2012. *A framework for K–12 science education: Practices, crosscutting concepts, and core ideas.* Washington, DC: National Academies Press.

What Are the Signs of Global Warming?

Scientists talk about the evidence that supports their concern that our planet is warming. Put an X next to all the statements that can be considered signs of global warming.

_____ **A.** The concentration of CO_2 is at 400 parts per million (ppm) today. It has never been more than 300 ppm in the past 800,000 years.

_____ **B.** Global sea level rose about 6.7 in. in the past century. The rate in the past decade, however, is nearly double that of the past century.

_____ **C.** Since 1880, the 10 hottest years on record have been in the past 17 years.

_____ **D.** Antarctica lost about 36 cubic miles of ice between 2002 and 2005.

_____ **E.** Since the beginning of the Industrial Revolution in the 1880s, the acidity of surface ocean waters has increased by about 30 percent.

_____ **F.** The area covered by sea ice in the Arctic at the end of summer has shrunk by about 40% since 1979.

_____ **G.** Currently, 37 glaciers in Glacier National Park are retreating.

_____ **H.** Since 1990, data show that birds are beginning their northern migration earlier and earlier each year—particularly birds that migrate over shorter distances. They are having trouble finding their normal food sources at their destination.

_____ **I.** The amount of heavy downpours has increased 74% in New England from 1958 to 2011.

Explain your thinking. What rule or reasoning did you use to decide if a statement is a sign that Earth is warming?

_____

_____

_____

What Are the Signs of Global Warming?

Teacher Notes

Purpose

The purpose of this assessment probe is to elicit students' ideas about signs of global warming. The probe is designed to determine whether students think a statement is accurate or complete enough to see direct patterns or draw inferences from data that can be used to support the claim that our planet is warming, or whether more long-term data is needed to decide whether the effect is a sign of global warming.

Type of Probe

Justified list

Related Concepts

Climate, climate change, evidence, global warming, weather

Explanation

The best answer is A, B, C, E, F, H, and I. A warming planet will have a number of effects, both direct and indirect, which are signs of global warming. Some of those

effects include average temperature increases, severe weather effects (e.g., droughts, floods, heavy rains, hurricanes), sea level rise, ocean warming, ocean acidification, glacial retreat, and animal migration pattern disruption. A warming planet can also cause the climate to change. Scientists define *climate* as the average weather in a place over at least 30 years. Whereas the weather can change in just a few hours or days, climate can take decades, hundreds, thousands, even millions of years to change. A number of indicators are used to determine if the planet is warming. This probe addresses some of them.

Answer choice A shows a marked increase in the concentration of carbon dioxide (CO_2) in the atmosphere over a long period of time. A student may know that CO_2 is a greenhouse gas and that when it increases in concentration, the planet-warming properties of the atmosphere increase as well. They may also have heard that an increase in CO_2 concentration in the atmosphere is a sign of global

warming, but not understand the greenhouse gas correlation. Their explanations should reveal their level of understanding.

Answer choices B, E, F, H, and I are indirect results of a warming planet. For answer B, when the ocean warms, it expands, causing some sea level rise. Other increases can be attributed to runoff from melting sea ice and glaciers worldwide. These measurements were made over a long period of time. For answer E, the increased levels of CO_2 in the atmosphere are causing the ocean to become more acidic as wave action dissolves CO_2 into the ocean water. Answer F is an inference that a warming planet would include ocean warming. The data have been collected for more than three decades. Answer H is an inference that the early arrival of migrating birds indicates that a warming planet is causing spring to arrive earlier than normal. For answer I, warm air holds more moisture than cooler air. As the planet warms, the atmosphere is able to hold more moisture, causing an increase in heavy downpours.

Answer choice C shows a direct measurement of increased temperatures since 1880.

Although answer choices D and G may be indicators of global warming, the data in D has not been collected over a long enough time period to attribute it to a warming planet or climate change. The same is true for answer choice G, which is a statement of current conditions. It does not include information on the time frame being studied, nor does it include data about other glaciers around the world. Additional time-frame data is needed to attribute the effect to global warming.

Administering the Probe

This probe is best used with grades 6–12 to elicit and discuss students' ideas about global warming and the types of evidence needed to support the claim that our planet is warming. Make sure students understand that they are to select statements that can be used to support the idea of a warming planet. Some statements are direct measurements of a warming planet and others are inferences of effects from a warming planet (e.g., sea level rise, ocean acidification, glacial retreat, heavy downpours). Listen or look for indications that students are considering the time frame in which the effect takes place as evidence of global warming.

Related Core Ideas in *Benchmarks for Science Literacy* (AAAS 2009)

9–12 The Earth
- Greenhouse gases in the atmosphere, such as carbon dioxide and water vapor, are transparent to much of the incoming sunlight but not to the infrared light from the warmed surface of the Earth. When greenhouse gases increase, more thermal energy is trapped in the atmosphere, and the temperature of the Earth increases the light energy radiated into space until it again equals the light energy absorbed from the Sun.

Related Core Ideas in *A Framework for K–12 Science Education* (NRC 2012)

6–8 ESS3.D: Global Climate Change
- Human activities, such as the release of greenhouse gases from burning fossil fuels, are major factors in the current rise in Earth's mean surface temperature (global warming). Reducing the level of climate change and reducing human vulnerability to whatever climate changes do occur depend on the understanding of climate science, engineering capabilities, and other kinds of knowledge, such as understanding of human behavior and on applying that knowledge wisely in decisions and activities.

9–12 ESS3.D: Global Climate Change

- Through computer simulations and other studies, important discoveries are still being made about how the ocean, the atmosphere, and the biosphere interact and are modified in response to human activities.

Related Research

- Researchers have found some scientifically acceptable ideas about global warming already present among 11-year-old students. Those ideas included the notion that an increase in the greenhouse effect will cause changes in weather patterns (Driver et al. 1994).
- In a study conducted by the Yale Project on Climate Change Communication, 80% of middle and high school students thought it was true, or mostly true, that climate changes from year to year (Leiserowitz, Smith, and Marlon 2011).
- Students have difficulty linking relevant elements of knowledge when explaining the greenhouse effect and may confuse the natural greenhouse effect with the enhancements of that effect (AAAS 2007; Andersson and Wallin 2000).

Suggestions for Instruction and Assessment

- This probe can be used with the formative assessment card sort strategy (Keeley 2008, 2015). Print each of the statements on cards and have students work in small groups to sort them into signs of global warming or not signs of global warming. During the card sort, have students justify their reasoning. Listen carefully for evidence that students can distinguish between direct and indirect signs that have long-term data versus an effect in which there is no comparison to longer term data. After instructional experiences have been provided to help students distinguish between signs that are considered evidence of global warming and signs that

may need more data, provide them with an opportunity to re-sort the cards and explain how their thinking changed.

- Have students sort their answer choices into statements that describe direct measurements of a warming Earth and statements that are inferences from data.
- Students can be easily misled by incomplete evidence. With instruction, a major goal of understanding global warming and climate change is for students to develop an awareness of claims and evidence that are complete, sufficient, and accurate.
- Students can learn more about 10 indicators of a warming world at NOAA's website: *www.cpo.noaa.gov/warmingworld*.
- The Climate Generation has published six interdisciplinary lesson plans that help students master the requisite background information on global climate change processes, the importance of the Arctic to global climate, and the potential effects of global warming in the Arctic. They also help students consider what could or should be done in response. Their website is available at *www.climategen.org/what-we-do/education/climate-change-and-energy-cirricula*.

References

American Association for the Advancement of Science (AAAS). 2007. Weather and climate. In *Atlas of science literacy*. Vol. 2., ed. AAAS, 20–21. Washington, DC: AAAS.

American Association for the Advancement of Science (AAAS). 2009. Benchmarks for science literacy online. *www.project2061.org/publications/bsl/online*.

Andersson, B., and A. Wallin. 2000. Students' understanding of the greenhouse effect, the societal consequences of reducing CO_2 emissions and the problem of ozone layer depletion. *Journal of Research in Science Teaching* 37 (10): 1096–1111.

Driver, R., A. Squires, P. Rushworth, and V. Wood-Robinson. 1994. *Making sense of secondary*

science: Research into children's ideas. London: Routledge.

Keeley, P. 2008. *Science formative assessment: 75 practical strategies for linking assessment, instruction, and learning.* Thousand Oaks, CA: Corwin Press and Arlington, VA: NSTA Press.

Keeley, P. 2015. *Science formative assessment: 75 practical strategies for linking assessment, instruction, and learning.* 2nd ed. Thousand Oaks, CA: Corwin Press.

Leiserowitz, A., Smith, N., and Marlon, J. R. 2011. American teens' knowledge of climate change. Yale Project on Climate Change Communication, Yale University. *http://environment.yale.edu/climate-communication/article/american-teens-knowledge-of-climate-change.*

National Research Council (NRC). 2012. *A framework for K–12 science education: Practices, crosscutting concepts, and core ideas.* Washington, DC: National Academies Press.

Section 3

Earth History, Weathering and Erosion, and Plate Tectonics

Concept Matrix.. 84

Related *Next Generation Science Standards* **Performance Expectations** .. 85

Related NSTA Resources 85

16 **How Old Is Earth?**.................................. 87
17 **Is It a Fossil?** 91
18 **Sedimentary Rock Layers** 95
19 **Is It Erosion?**... 99
20 **Can a Plant Break Rocks?**.....................103
21 **Grand Canyon**....................................... 107
22 **Mountains and Beaches** 111
23 **How Do Rivers Form?** 117
24 **What Is the Inside of Earth Like?**...........121
25 **Describing Earth's Plates** 125
26 **Where Do You Find Earth's Plates?** 131
27 **What Do You Know About Volcanoes and Earthquakes?**.................135

Concept Matrix
Probes #16–#27

PROBES	#16 How Old Is Earth?	#17 Is It a Fossil?	#18 Sedimentary Rock Layers	#19 Is It Erosion?	#20 Can a Plant Break Rocks?	#21 Grand Canyon	#22 Mountains and Beaches	#23 How Do Rivers Form?	#24 What Is the Inside of Earth Like?	#25 Describing Earth's Plates	#26 Where Do You Find Earth's Plates?	#27 What Do You Know About Volcanoes and Earthquakes?
GRADE RANGE →	6–12	3–8	6–12	3–12	3–5	3–8	5–12	3–12	6–12	5–12	6–12	3–12
CORE CONCEPTS ↓												
age of Earth	X											
canyons						X						
continental crust											X	
deposition				X			X					
Earth's crust										X		
Earth history	X	X										
earthquakes										X		X
erosion				X		X	X	X				
fossil		X										
geologic time	X		X									
inner and outer core									X			
landforms						X	X					
mantle									X			
mountains							X					
ocean basin crust											X	
plate tectonics										X	X	X
principle of superposition			X									
radiometric dating	X											
river systems						X		X				
rock layers			X									
sedimentary rock			X									
structure of Earth									X	X		
tributaries								X				
volcanoes											X	X
weathering				X	X	X	X					

Related *Next Generation Science Standards* Performance Expectations (NGSS Lead States 2013)

· ·

Earth Systems

- Kindergarten, K-ESS2-2: Construct an argument supported by evidence for how plants and animals (including humans) can change the environment to meet their needs.

- Grade 2, 2-ESS1-1: Use information from several sources to provide evidence that Earth events can occur quickly or slowly.

- Grade 4, 4-ESS1-1: Identify evidence from patterns in rock formations and fossils in rock layers to support an explanation for changes in a landscape over time.

- Grade 4, 4-ESS2-1: Make observations and/or measurements to provide evidence of the effects of weathering or the rate of erosion by water, ice, wind, or vegetation.

- Grades 9–12, HS-ESS2-3: Develop a model based on evidence of Earth's interior to describe the cycling of matter by thermal convection.

Earth History

- Grades 6–8, MS-ESS1-4: Construct a scientific explanation based on evidence from rock strata for how the geologic time scale is used to organize Earth's 4.6-billion-year-old history.

- Grades 6–8, MS-ESS2-1: Develop a model to describe the cycling of Earth's materials and the flow of energy that drives this process.

- Grades 6–8, MS-ESS2-2: Construct an explanation based on evidence for how geoscience processes have changed Earth's surface at varying time and spatial scales.

- Grades 6–8, MS-ESS2-3: Analyze and interpret data on the distribution of fossils and rocks, continental shapes, and seafloor structures to provide evidence of the past plate motions.

- Grades 9–12, HS-ESS1-6: Apply scientific reasoning and evidence from ancient Earth materials, meteorites, and other planetary surfaces to construct an account of Earth's formation and early history.

- Grades 9–12, HS-ESS2-1: Develop a model to illustrate how Earth's internal and surface processes operate at different spatial and temporal scales to form continental and ocean-floor features.

Reference

NGSS Lead States. 2013. *Next Generation Science Standards: For states, by states.* Washington, DC: National Academies Press. *www.nextgenscience. org/next-generation-science-standards.*

Related NSTA Resources

NSTA Press Books

Environmental Literacy Council, and National Science Teachers Association. 2007. *Earthquakes, volcanoes, and tsunamis: Resources for environmental literacy.* Arlington, VA: NSTA Press.

Fullager, P., and N. West. 2011. *Project Earth science: Geology.* 2nd ed. Arlington, VA: NSTA Press.

Kastens, K., and M. Turrin. 2010. *Earth science puzzles: Making meaning from data.* Arlington, VA: NSTA Press.

Konicek-Moran, R. 2013. *Everyday Earth and space science mysteries: Stories for inquiry-based science teaching.* Arlington, VA: NSTA Press.

NSTA Journal Articles

Blank, L., M. Plautz, H. Almquist, J. Crews, and J. Estrada. 2012. Using Google Earth to teach plate tectonics and science explanations. *Science Scope* 35 (9): 41–48.

Clary, R., and J. Wandersee. 2008. How old? Tested and trouble-free ways to convey geologic time. *Science Scope* 33 (4): 62–66.

Coffey, P., and S. Mattox. 2006. Take a tumble: Weathering and erosion using a rock tumbler. *Science Scope* 29 (6): 33–36.

Ford, B., and M. Taylor. 2006. Investigating students' ideas about plate tectonics. *Science Scope* 30 (1): 38–43.

Hester, P. 2008. Science sampler: Taking steps to understand geologic time. *Science Scope* 32 (2): 54–56.

Mulvey, B., and R. Bell. 2012. A virtual tour of plate tectonics. *The Science Teacher* 79 (6): 53–58.

Norell, M. 2003. Science 101: What is a fossil? *Science and Children* 41 (6): 20.

Robertson, B. 2103. Science 101: How do we figure out what happened to the Earth in the past? *Science and Children* 50 (8): 76–79.

Schipper, S., and S. Mattox. 2010. Using Google Earth to study the basic characteristics of volcanoes. *Science Scope* 34 (3): 28–37.

Schomburg, A. 2003. Real earthquakes, real learning. *Science and Children* 41 (1): 26–30.

Weinburg, M., and C. Silva. 2011. Math, science, and models. *Science and Children* 49 (1): 58–62.

NSTA Learning Center Resources
NSTA Webinars

NGSS Core Ideas: Earth's Systems
http://learningcenter.nsta.org/products/ symposia_seminars/NGSS/webseminar32.aspx
Plate Tectonics Made to Order
http://learningcenter.nsta.org/products/ symposia_seminars/NSDL/webseminar4.aspx

NSTA Science Objects

Earth's Changing Surface
http://learningcenter.nsta.org/ resource/?id=10.2505/6/SCP-ECS.0.1
Earth's Changing Surface: Changing Earth From Within
http://learningcenter.nsta.org/ resource/?id=10.2505/7/SCB-ECS.1.1
Earth's Changing Surface: Sculpting the Landscape
http://learningcenter.nsta.org/ resource/?id=10.2505/7/SCB-ECS.2.1
Plate Tectonics: Layered Earth
http://learningcenter.nsta.org/ resource/?id=10.2505/7/SCB-PT.1.1
Plate Tectonics: Lines of Evidence
http://learningcenter.nsta.org/ resource/?id=10.2505/7/SCB-PT.5.1
Plate Tectonics: Plate Interactions
http://learningcenter.nsta.org/ resource/?id=10.2505/7/SCB-PT.3.1
Plate Tectonics: Plates
http://learningcenter.nsta.org/ resource/?id=10.2505/7/SCB-PT.2.1
Rocks: Earth's Autobiography
http://learningcenter.nsta.org/ resource/?id=10.2505/7/SCB-RK.4.1

How Old Is Earth?

Six friends made a birthday card for Earth to celebrate Earth Day. Each friend had a different estimate for the age of Earth. This is what they said:

Velma: Earth's age is less than 10,000 years.

Kari: Earth's age is between 10,000 years and 1 million years.

Jonah: Earth's age is between 1 million and 1 billion years.

Dottie: Earth's age is between 1 billion and 10 billion years.

Ling: Earth's age is between 10 billion and 1 trillion years.

Matt: Earth's age is more than 1 trillion years.

Which friend do you agree with the most? _____ Explain your thinking. Include a number that is your best estimate of the age of Earth.

_____

_____

_____

_____

How Old Is Earth?

Teacher Notes

Purpose

The purpose of this assessment probe is to elicit students' ideas about Earth's history. The probe is designed to reveal how old students think Earth is.

Type of Probe

Friendly talk

Related Concepts

Age of Earth, Earth history, geologic time, radiometric dating

Explanation

The best answer is Dottie's: "Earth's age is between 1 billion and 10 billion years." Although the exact age of Earth cannot be determined directly, the best estimate is based on evidence from radiometric dating of various kinds of rock, both on Earth and beyond Earth, to estimate Earth's age. For example, scientists search for and date the oldest rocks exposed on Earth's surface.

Scientists also use radiometric dating to determine the ages of meteorites. *Meteorites* are space rocks that once orbited the Sun during the formation of the solar system (of which Earth is part of and was forming), but later entered Earth's atmosphere and fell to Earth (and continue to fall to Earth). Likewise, scientists use radiometric dating to determine the ages of moon rocks obtained by astronauts. When all these data are taken together, they suggest that the age of Earth, meteorites, the Moon—and by inference our entire solar system—is somewhere between 4.5 billion to 4.6 billion years old.

Administering the Probe

This probe is best used with students in grades 6–12 who can grasp the magnitude of large numbers. Encourage students to include in their explanation the type of evidence they think scientists use to support their ideas about the age of Earth. Because the answer choices are given in broad ranges, ask students

 **16**

to give an estimated age, instead of a range, in their explanation.

Related Core Ideas in *Benchmarks for Science Literacy* (AAAS 2009)

9–12 The Universe

- Our solar system coalesced out of a giant cloud of gas and debris left in the wake of exploding stars about 5 billion years ago. Everything in and on the Earth, including living organisms, is made of this material.

9–12 Processes That Shape the Earth

- Scientific evidence indicates that some rock layers are several billion years old.

9–12 Extending Time

- Prior to the 1700s, many considered the Earth to be just a few thousand years old. By the 1800s, scientists were starting to realize that the Earth was much older even though they could not determine its exact age.
- In the early 1800s, Charles Lyell argued in *Principles of Geology* that the Earth was vastly older than most people believed. He supported his claim with a wealth of observations of the patterns of rock layers in mountains and the locations of various kinds of fossils.

Related Core Ideas in *A Framework for K–12 Science Education* (NRC 2012)

6–8 ESS1.C: History of the Earth

- The geologic time scale interpreted from rock strata provides a way to organize Earth's history. Analyses of rock strata and the fossil record provide only relative dates, not an absolute scale.

9–12 ESS1.C: History of the Earth

- Continental rocks, which can be older than 4 billion years, are generally much

older than the rocks of the ocean floor, which are less than 200 million years old.

- Although active geologic processes, such as plate tectonics and erosion, have destroyed or altered most of the early rock record on Earth, other objects in the solar system, such as lunar rocks, asteroids, and meteorites, have changed little over billions of years. Studying these objects can provide information about Earth's formation and early history.

Related Research

- An understanding of geologic time may be hindered by students' ability to perceive the relative magnitude of large numbers and, hence, large periods of time (Cheek 2012).
- High school and college students who understand Earth's geological age—about 4.5 billion years—are much more likely to understand evolution (Cotner, Brooks, and Moore 2010).
- A recent study by Jones, Taylor, and Broadwell (2009) reported that visually impaired students performed better than their sighted peers when using extremely large scales, such as geologic time. They concluded that this may have significant considerations for the classroom because most teachers use visual analogies to convey large spans of time. The researchers suggest that auditory and kinesthetic analogies may provide better learning opportunities for students.

Suggestions for Instruction and Assessment

- Some students may be influenced by biblical interpretation, which implies Earth is 6,000 years old. These are religious beliefs, not scientific facts based on empirical evidence. Be aware that if students search the internet for information about the age of Earth, they are likely to discover faith-based sites.

Make sure students can distinguish scientifically based information on the internet from religion-based information. Teaching the nature of science in conjunction with Earth history can help students understand how scientists estimate the age of Earth.

- To understand the age of Earth in billions of years, students need to have conceptual awareness of just how large 1 billion is. One way to visualize the difference between 1 million and 1 billion is to draw a 10 cm line across a piece of paper. Tell students the line represents 1 billion. Ask them to put a mark where 1 million is. (Many students will mark the middle of the line.) In actuality, the mark would be barely past the beginning of the line.

- The historical sequence of changing ideas about the age of Earth can be a fascinating story to show students how science progresses when new knowledge is gained. Like early religious beliefs that the Sun revolved around Earth, which later changed, the age of Earth also raised questions about religious beliefs versus scientific evidence. The change in the conception of the age of Earth—from a few thousand to many millions of years—proposed by scientists in the 1800s was dramatic and, for most people, beyond belief. The estimated age was unimaginably greater than the prevailing beliefs. It was also based on the assumption that Earth's features (mountains, valleys, etc.) had been formed gradually by processes still underway, not in a single, instantaneous creation (AAAS 2009). Contrast this with the indirect evidence and radiometric data from rock layers, meteorites, and lunar rocks that scientists use today to estimate the age of Earth in billions of years.

- Have students explore how the early Earth was very different, in terms of landforms and atmosphere, from what it is like today.

- Emphasis at the high school level should be on using available evidence within the solar system to reconstruct the early history of Earth, which formed along with the rest of the solar system about 4.6 billion years ago. Examples of evidence they can use include the absolute ages of ancient materials (obtained by radiometric dating of meteorites, moon rocks, and Earth's oldest minerals), the sizes and compositions of solar system objects, and the impact cratering record of planetary surfaces.

References

American Association for the Advancement of Science (AAAS). 2009. Benchmarks for science literacy online. *www.project2061.org/publications/bsl/online.*

Cheek, K. 2012. Students' understanding of large numbers as a key factor in their understanding of geologic time. *International Journal of Science and Mathematics Education* 10 (5): 1047–1069.

Cotner, S, D. C. Brooks, and R. Moore. 2010. Is the age of the Earth one of our "sorest troubles"? Students' perceptions about deep time affect their acceptance of evolutionary theory. *Evolution* 64 (3): 858–864.

Jones, M., A. Taylor, and B. Broadwell. 2009. Concepts of scale held by students with visual impairment. *Journal of Research in Science Teaching* 46 (5): 506–519.

National Research Council (NRC). 2012. *A framework for K–12 science education: Practices, crosscutting concepts, and core ideas.* Washington, DC: National Academies Press.

Is It a Fossil?

Scientists use fossils to study the past. Put an X next to the things that are considered fossils.

_____ **A.** dinosaur bone

_____ **B.** mammoth frozen in ice

_____ **C.** dinosaur tracks

_____ **D.** sedimentary rock

_____ **E.** ancient shark tooth

_____ **F.** Egyptian mummy

_____ **G.** air trapped in an ancient ice layer

_____ **H.** dinosaur excrement ("poop")

_____ **I.** prehistoric cave drawing (petroglyph)

_____ **J.** Roman skull from 100 BC

_____ **K.** insect trapped in amber

_____ **L.** prehistoric tool

_____ **M.** petrified wood

_____ **N.** hardened lava

_____ **O.** a 1 billion-year-old rock

_____ **P.** cast of a shell in rock

_____ **Q.** leaf imprint in a rock

_____ **R.** old carved gravestone

_____ **S.** piece of dinosaur eggshell

_____ **T.** ancient arrowhead

Explain your thinking. What rule or reasoning did you use to decide if something was a fossil?

_____

_____

_____

_____

Is It a Fossil?

Teacher Notes

Purpose

The purpose of this assessment probe is to elicit students' ideas about fossils. The probe is designed to find out if students distinguish fossils from other ancient things or whether they think only bones and shells form fossils.

Type of Probe

Justified list

Related Concepts

Earth history, fossil

Explanation

The best answer is A, B, C, E, H, K, M, P, Q, S. The things that are not considered fossils are choices D, F, G, I, J, L, N, O, R, and T. A *fossil* is the naturally preserved remains or trace of an organism that lived in the geologic past. These remains or traces are usually preserved in sediments that hardened and formed rock over time, or the body parts themselves became hardened as minerals replaced soft body tissue.

Paleontologists date fossils as being at least 10,000 years old. When people think of fossils, they typically think of hard body parts such as bones or shells, or imprints of the organism's body in rock. Whole organisms, including their flesh, can also be fossils, such as wooly mammoths frozen in ice and ancient insects trapped in amber. Traces of ancient organisms are another type of fossil. These include traces of a once-living organism such as footprints, egg shells, nests, and even waste droppings (e.g., dinosaur poop!).

Ancient human artifacts such as petroglyphs, arrowheads, and prehistoric tools are considered archeological artifacts rather than fossils, even though some are older than 10,000 years. Although rock often bears evidence of fossils, rocks by themselves are not considered fossils.

Administering the Probe

This probe is best used with grades 3–8. Take off any of the items that are unfamiliar

to your students or go through the list and explain what each one is, or use pictures if available. This probe can be combined with the card sort formative assessment classroom technique described on pages 3–4.

Related Core Ideas in *Benchmarks for Science Literacy* (AAAS 2009)

3–5 Evolution of Life
- Fossils can be compared to one another and to living organisms according to their similarities and differences. Some organisms that lived long ago are similar to existing organisms, but some are quite different.

6–8 Processes That Shape the Earth
- Sediments of sand and smaller particles (sometimes containing the remains of organisms) are gradually buried and are cemented together by dissolved minerals to form solid rock again.
- Thousands of layers of sedimentary rock confirm the long history of the changing surface of the Earth and the changing life forms whose remains are found in successive layers.

6–8 Evolution of Life
- Many thousands of layers of sedimentary rock provide evidence for the long history of the Earth and for the long history of changing life forms whose remains are found in the rocks.

Related Core Ideas in *A Framework for K–12 Science Education* (NRC 2012)

3–5 LS4.A: Evidence of Common Ancestry and Diversity
- Fossils provide evidence about the types of organisms that lived long ago and also about the nature of their environments.

6–8 ESS1.C: The History of Planet Earth
- The geologic time scale interpreted from rock strata provides a way to organize Earth's history. Analyses of rock strata and the fossil record provide only relative dates, not an absolute scale.

6–8 LS4.A: Evidence of Common Ancestry and Diversity
- The collection of fossils and their placement in chronological order (e.g., through the location of the sedimentary layers in which they are found or through radioactive dating) is known as the fossil record. It documents the existence, diversity, extinction, and change of many life forms throughout the history of life on Earth.

Related Research
- Oversby's study (1996) found that a majority of preservice teachers surveyed did not think that a footprint preserved in rock and a mammoth frozen in ice were fossils.

Suggestions for Instruction and Assessment
- Have students sort the examples of fossils listed on the probe by fossils of body parts and fossils that are traces of organisms but not body parts.
- Challenge older students to consider whether ancient DNA or pollen is considered to be a fossil.
- Challenge students to find out whether fossil fuels such as coal and petroleum are considered fossils.
- Have students research the different ways fossils are formed.
- Show several pictures of mold fossils, cast fossils, trace fossils, and true form fossils. Have students group them into four types of fossils according to their own classification scheme.

References

American Association for the Advancement of Science (AAAS). 2009. Benchmarks for science literacy online. *www.project2061.org/publications/bsl/online.*

National Research Council (NRC). 2012. *A framework for K–12 science education: Practices,* *crosscutting concepts, and core ideas.* Washington, DC: National Academies Press.

Oversby, J. 1996. Knowledge of Earth science and the potential for its development. *School Science Review* 78 (283): 91–97.

Sedimentary Rock Layers

Five friends were standing at the foot of a tall, rocky cliff. They noticed several layers of sedimentary rock in the cliff. They each had different ideas about the ages of the layers of rock. This is what they said:

Herman: I think the youngest layers are always at the top.

Joyce: I think the youngest layers are always at the bottom.

Flora: I think the youngest layers are sometimes at the top.

Steve: I think the oldest layers are usually in the center.

Cliff: I think all rock layers are the same age; the layers are just different types of rock.

Who do you think has the best idea? _____ Explain your thinking.

_____

_____

_____

_____

Sedimentary Rock Layers

Teacher Notes

Purpose

The purpose of this assessment probe is to elicit students' ideas about layers of sedimentary rock. The probe is designed to find out how students think about the age of rock layers.

Type of Probe

Justified list

Related Concepts

Geologic time, principle of superposition, rock layers, sedimentary rock

Explanation

The best answer is Flora's: "I think the youngest layers are sometimes at the top." Younger rock layers being at the top is generally true for sedimentary rock layers. Layers of sedimentary rock are the result of a series of depositional events, such as sedimentation. Sedimentary layers are deposited following the principle of superposition. The layers formed one after the other with each added layer at the top being

younger than the ones below. Relative ages of rock layers being oldest at the bottom and gradually younger as you move to the next layer above is generally true for sedimentary rock. However, sometimes folding, faulting, and uplift may change the order of the layers so that the ones on top are not always the youngest.

Administering the Probe

This probe is best used with grades 6–12. Students should be familiar with sedimentary rocks before using this probe. Make sure students know that the probe is about layers of sedimentary rock, not rock in general.

Related Core Ideas in *Benchmarks for Science Literacy* (AAAS 2009)

6–8 Processes That Shape the Earth

- Thousands of layers of sedimentary rock confirm the long history of the changing

surface of the Earth and the changing life forms whose remains are found in successive layers. The youngest layers are not always found on top, because of folding, breaking, and uplift of layers.

9–12 Processes That Shape the Earth

- Scientific evidence indicates that some rock layers are several billion years old.

Related Core Ideas in *A Framework for K–12 Science Education* (NRC 2012)

3–5 ESS1.C: The History of Planet Earth

- Local, regional, and global patterns of rock formations reveal changes over time due to Earth forces, such as earthquakes. The presence and location of certain fossil types indicate the order in which rock layers were formed.

6–8 ESS1.C: The History of Planet Earth

- The geologic time scale interpreted from rock strata provides a way to organize Earth's history. Analyses of rock strata and the fossil record provide only relative dates, not an absolute scale.

Related Research

- Ault (1982) probed children's conceptions of geologic time. He found that children who were able to accurately complete a sequencing task with a hypothetical compost pile were unable to transfer those ideas to rock outcrops in the local area. Some children in the sample held to an accretionary view of rock formation and said that the oldest rock layers were in the center.
- Few children in a study by Happs (1982) recognized the relationship between sedimentary rocks and the sedimentary processes by which they are formed.
- Some students may hold the view that the world has always been the same as it is now, or that any changes that have occurred

must have been sudden and comprehensive. This view may account for some students thinking the rock layers are all the same age (Freyberg 1985).

Suggestions for Instruction and Assessment

- Take students on a field trip to observe a rock outcrop, or road side cut, and describe the layers. (Safety notes: Follow all school and district safety procedures for field trips. Use caution when walking near rock outcrops because sharp edges from the rock formations and debris can cut skin or puncture footwear.)
- Have students examine and describe online images of sedimentary rock layers.
- Help students understand the concept of relative layers.
- Have students investigate how scientists use rock layers to establish the geologic time scale. The American Museum of Natural History offers an activity called "Solve a Sedimentary Layers Puzzle" to help students understand how fossils are used to determine if rock layers are from the same geologic time period. The activity is available at *www.amnh.org/explore/curriculum-collections/ dinosaurs-ancient-fossils-new-discoveries/ solve-a-sedimentary-layers-puzzle*.
- Students can use a model to see how sedimentary rocks form layers. Provide samples of soil, or have students bring soil in from their local areas. Have students examine the soil samples with hand lenses and describe them. They can put an inch of soil in a clear 12 oz. cup. Have them fill the cup with water and stir vigorously. The larger particles in the soil will settle to the bottom, with the successively smaller particles forming layers (gravel, sand, silt, clay). They can imagine that successive rain storms washed all these particles into a lakebed; with time and pressure, the

layers they see can form sedimentary rocks. (Safety notes: Make sure soil samples are from pesticide- and herbicide-free sources. Caution students about possible sharp materials in the soil that could cut skin. Immediately wipe up any spilled water to avoid slips and falls. Be sure students wash their hands with soap and water after handling soil.)

References

American Association for the Advancement of Science (AAAS). 2009. Benchmarks for science literacy online. *www.project2061.org/publications/bsl/online.*

Ault, C. 1982. Time in geological explanations as perceived by elementary-school students. *Journal of Geological Education* 30 (5): 304–309.

Freyberg, P. 1985. Implications across the curriculum. In *Learning in science: The implications of children's science,* ed. R. Osborne and P. Freyberg, 125–135. Auckland, New Zealand: Heinemann.

Happs, J. 1982. Some aspects of student understanding of rocks and minerals. Science Education Research Unit Working Paper 204, University of Waikato, Hamilton, New Zealand.

National Research Council (NRC). 2012. *A framework for K–12 science education: Practices, crosscutting concepts, and core ideas.* Washington, DC: National Academies Press.

Is It Erosion?

Erosion is a process that changes the surface of Earth. Put an X next to the changes you think are examples of erosion.

_____ **A.** soil from a riverbank being carried downstream

_____ **B.** rock steps worn down from millions of visitors

_____ **C.** a rock breaking into smaller pieces when water freezes in the cracks

_____ **D.** a huge boulder transported by a glacier

_____ **E.** mud sliding down a hill during a landslide

_____ **F.** sand deposited on a beach from the ocean

_____ **G.** a tunnel formed in soil by a burrowing animal

_____ **H.** a rock shattered by a large rock that fell on it

_____ **I.** soil from a cliff being washed into the ocean

_____ **J.** fragments of rock sliding down a mountain cliff

_____ **K.** a sharp rock in a stream slowly getting smoothed out and rounded

_____ **L.** desert sand being carried by the wind

_____ **M.** plant roots expanding a crack in a rock

_____ **N.** the date on a 200-year-old gravestone worn away

Explain your thinking. What rule or reasoning did you use to decide which processes are examples of erosion?

_____

_____

_____

_____

Is It Erosion?

Teacher Notes

Purpose

The purpose of this assessment probe is to elicit students' ideas about erosion. The probe is designed to find out whether students can distinguish between erosion and weathering.

Type of Probe

Justified list

Related Concepts

Deposition, erosion, weathering

Explanation

The best answer is A, D, E, I, J, and L. The processes of weathering and erosion are often confused at all grade levels, even after instruction. *Erosion* is the transfer of earth material, such as rock, soil, mud, or weathered sediment, to a new location. Wind, water, and ice are examples of erosion agents that carry the material. Erosion also happens in response to gravity (materials sliding or falling downward), or as a result of human activities such

as deforestation. Answers B, C, H, K, M, and N are examples of weathering. *Weathering* is the breaking up and wearing down of rock material. Weathering can be caused by water, ice, air, chemicals, heat, pressure, or living things. Answer F is an example of deposition. After rock has weathered into small pieces carried by rivers (erosion) into the ocean, these tiny pieces may be deposited on a beach as sand. Answer choice G involves an animal digging through the soil, but the soil is not carried away to a new location. Weathering and erosion are often confused. When rock material is broken down or changed, but not transported to a new location, the process is called *weathering*. When the soil or broken down rock material is moved away and carried to a new location, the process is called *erosion*.

Administering the Probe

This probe can be used with grades 3–12. This probe can be used with the card sort strategy described on pages 3–4.

Related Core Ideas in *Benchmarks for Science Literacy* (AAAS 2009)

3–5 Processes That Shape the Earth

- Waves, wind, water, and ice shape and reshape the Earth's land surface by eroding rock and soil in some areas and depositing them in other areas, sometimes in seasonal layers.

6–8 Processes That Shape the Earth

- The Earth's surface is shaped in part by the motion of water (including ice) and wind over very long times, which acts to level mountain ranges. Rivers and glacial ice carry off soil and break down rock, eventually depositing the material in sediments or carrying it in solution to the sea.

Related Core Ideas in *A Framework for K–12 Science Education* (NRC 2012)

3–5 ESS2.A: Earth's Systems

- Rainfall helps to shape the land and affects the types of living things found in a region. Water, ice, wind, living organisms, and gravity break rocks, soils, and sediments into smaller particles and move them around.

Related Research

- In a survey of 236 students ages 16–19, many of the students regarded weathering as solely related to atmospheric elements such as rain and wind. Human actions were perceived as types of erosion accelerated by humans. Students were not sure whether animal activities were examples of bioerosion or biological weathering (Dove 1997).
- When children ages 7–11 years old were shown a picture of a weathered gravestone, most were able to recognize that it was worn away by a natural process, which mountains and cliffs also experience, but

students inconsistently understood that the process would continue to wear away the gravestone and sometimes were able to cite a cause. Some of these children thought that age was the causal agent that wore away the gravestone (Blake 2005).
- A literature review revealed that misuse of terminology is an issue when describing Earth processes. Even college students confuse the terms *weathering* and *erosion* (Cheek 2010).

Suggestions for Instruction and Assessment

- This probe can be turned into a card sort activity (Keeley 2015). Have the students sort the cards into two vertical columns: (1) examples of erosion and (2) non-examples of erosion. Have students explain their reasons for sorting the examples the way they did. In the process, come up with an operational definition of erosion.
- For the non-examples of erosion, ask students to describe what is different about the processes. Use their ideas to develop the concepts of weathering and deposition.
- After students have learned about weathering, erosion, and deposition, have them sort the examples into those three processes.
- Help students be aware that the word *erosion* has a different meaning in our everyday language. The colloquial meaning of erosion is "wearing away," rather than being "carried away." This colloquial definition may be one of the reasons students have difficulty distinguishing between the two processes.
- Have students develop a model that can be used to explain the difference between weathering, erosion, and deposition.
- Have students look for examples of erosion in their local area. If possible, take photos and use them to explain how erosion changed the area.
- Use stream tables to model the process of erosion.

- Search online for images of weathering and erosion of landforms and use the images to have students describe what caused the changes in the landform over time.

References

American Association for the Advancement of Science (AAAS). 2009. *Benchmarks for science literacy online. www.project2061.org/publications/bsl/online.*

Blake, A. 2005. Do young children's ideas about the Earth's structure and processes reveal underlying patterns of descriptive and causal understanding in Earth science? *Research in Science and Technological Education* 23 (1): 59–74.

Cheek, K. 2010. Commentary: A summary and analysis of twenty-seven years of geosciences conceptions research. *Journal of Geoscience Education* 58 (3): 122–134.

Dove, J. 1997. Student ideas about weathering and erosion. *International Journal of Science Education* 19 (8): 971–980.

Keeley, P. 2015. *Science formative assessment: 75 practical strategies for linking assessment, instruction, and learning.* 2nd ed. Thousand Oaks, CA: Corwin Press.

National Research Council (NRC). 2012. *A framework for K–12 science education: Practices, crosscutting concepts, and core ideas.* Washington, DC: National Academies Press.

Can a Plant Break Rocks?

Four friends were talking about rocks. They each had different ideas about what could break up rocks. This is what they said:

Luisa: I think plants can break up rocks.

Ivan: I agree with Luisa, but they would have to be large plants like trees.

Sam: I disagree with both of you. Plants can't break up a rock. Water, ice, and wind break up rocks.

Mary: I disagree with all of you. To break up a rock, something hard has to fall on it or crack it.

Who do you agree with the most? _____ Explain why you agree.

_____

_____

_____

_____

Can a Plant Break Rocks?

Teacher Notes

Purpose

The purpose of this assessment probe is to elicit students' ideas about weathering. The probe is designed to find out whether students recognize plants, including small plants, can cause weathering of rock.

Type of Probe

Friendly talk

Related Concepts

Weathering

Explanation

The best answer is Luisa's: "I think plants can break up rocks." Plant roots, even those of tiny plants, exert stress or substantial pressure on rock as they grow larger, causing rock to crack and even break apart through mechanical weathering. Plants can also break down rock through chemical weathering by releasing acid, which dissolves parts of the rock. Sam is partially correct. Water, ice, and wind can also break away pieces of rock. However, Sam is incorrect in saying plants can't break up rocks. Mary is also partially correct. Rock does break up when other rocks fall on it or humans break rocks. However, Mary does not recognize the other ways rock can weather.

Administering the Probe

This probe is best used with grades 3–5. Make sure students understand that when picking the person they most agree with, they must agree with the entire statement, not just a part of it. Their explanation can include parts of other statements, but they should justify why they agree more with one person than the others. This probe can be combined with the four corners formative assessment classroom technique discussed in the Suggestions for Instruction and Assessment section.

Related Core Ideas in *Benchmarks for Science Literacy* (AAAS 2009)

K–2 Processes That Shape the Earth

- Animals and plants sometimes cause changes in their surroundings.

3–5 Processes That Shape the Earth

- Rock is composed of different combinations of minerals. Smaller rocks come from the breakage and weathering of bedrock and larger rocks.

Related Core Ideas in *A Framework for K–12 Science Education* (NRC 2012)

K–2 ESS2.E: Biogeology

- Plants and animals can change their environment.

3–5 ESS2.A: Earth Materials and Systems

- Rainfall helps to shape the land and affects the types of living things found in a region. Water, ice, wind, living organisms, and gravity break rocks, soils, and sediments into smaller particles and move them around.

Related Research

- AAAS Project 2061 tested middle and high school students for common misconceptions and alternative ideas. A multiple choice item that revealed whether students understand that plant roots could break up rocks showed that 27% of the students who answered the item failed to recognize this (AAAS 2015).

Suggestions for Instruction and Assessment

- Conduct the four corners formative assessment classroom technique. Label four corners of the classroom with each answer choice. Students stand in the corner that represents their answer choice and discuss their ideas with others in the same corner.

After their discussion, each corner group presents their reasoning to support their answer choice.

- Ask students to generate examples of different ways rocks break apart into large and small pieces.
- Have students draw a sequence of pictures, starting with a seed that falls into a crack in a rock, then grows, and then splits the rock.
- Observe sidewalks and other paved surfaces around the schoolyard for evidence of plants growing in cracks and predict what might happen if the plant continues to grow. (Safety note: Do not handle any poisonous plants such as poison ivy, which can elicit an allergic reaction.)
- If students have smartphones or other devices with cameras, have them take pictures in their local area of plants growing in cracks of rock or rocky outcrops that expose tree roots. Have them bring their photos and use them to create a photographic display of evidence that shows that plants can break rocks.
- Have students create a model to show how roots can break a rock as they grow. Have them mix plaster of Paris and pour it into a disposable cup. Students should place several bean seeds just below the surface and place a few so that half the bean is above the surface. Then, they should cover the cup with a damp towel until the seeds sprout. Over time, students will see the plaster crack and flake away in places as the seedling grows. (Safety notes: Students should wear sanitized safety goggles, non-latex gloves, and aprons during the entire activity. Review and share safety instructions for plaster of Paris with students. Make sure there is appropriate ventilation when working with plaster of Paris powder. Be sure students wash their hands with soap and water after completing this activity.)

- Search for images online using the keywords "plant roots break rock" to obtain a variety of images that show rocks that are cracked or broken apart by a plant that grew in the rock.

References

American Association for the Advancement of Science (AAAS). 2009. Benchmarks for science literacy online. *www.project2061.org/publications/bsl/online.*

American Association for the Advancement of Science (AAAS). 2015. Question: Both the growth of plant roots and the freezing of water can break Earth's solid rock layer. Item WE007002, Project 2061, AAAS Science Assessment. *http://assessment.aaas.org/items/WE007002#/1.*

National Research Council (NRC). 2012. *A framework for K–12 science education: Practices, crosscutting concepts, and core ideas.* Washington, DC: National Academies Press.

Grand Canyon

Six friends were standing along the rim of the Grand Canyon. Looking down, they could see layers of rock and the Colorado River at the bottom. They wondered how the Grand Canyon formed. They each had a different idea. This is what they said:

Natara: I think the Grand Canyon was formed when Earth formed. It has just gotten bigger over time.

Cecil: I think the Grand Canyon formed from earthquakes that cracked open the land and pulled it apart.

Garth: I think the Colorado River and streams slowly carved out the Grand Canyon.

Robert: I think a huge flood rushed through the land and formed the Grand Canyon.

Kumiyo: I think the river got so heavy that it sunk down through the rock and formed the walls of the Grand Canyon.

Luna: I don't agree with any of your ideas. I think the Grand Canyon was formed in some other way.

Who do you think has the best idea? _____ Explain your thinking.

_____

_____

_____

Grand Canyon

Teacher Notes

Purpose

The purpose of this assessment probe is to elicit students' ideas about the formation of the Grand Canyon. The probe is designed to find out if students recognize the role of rivers and flowing water and connect them to the processes of weathering and erosion in the formation of canyons.

Type of Probe

Friendly talk

Related Concepts

Canyons, erosion, landforms, river systems, weathering

Explanation

The best answer is Garth's: "I think the Colorado River and streams slowly carved out the Grand Canyon." The oldest rocks in the Grand Canyon are over 1 billion years old. Although several factors contributed to

the formation of the Grand Canyon over a long period of time, the primary processes responsible for the canyon we see today are weathering and erosion by river systems. About 6 million years ago, the Colorado River began to cut through the upper rock layers. (Note: Although 6 million years is still the most widely accepted age for the onset of the carving of the Grand Canyon, some scientists have proposed that it may have begun around 16 million years ago). This cutting process happened slowly, inch by inch. The water, wind, and ice weathered away pieces of rock—from small grains to large boulders—which were carried away by the Colorado River. Smaller side streams and tributaries branched off the Colorado River and carved out other sections of the Grand Canyon. As the Colorado River moves rock and sediment downriver, it continues to scour the riverbed and carve away at the banks, thereby widening and deepening the river.

Administering the Probe

This probe is best used with grades 3–8. Before using this probe, have students describe what a canyon is and what it looks like. If students are not familiar with the Grand Canyon, show them images of the canyon (including the Colorado River and side streams) before they answer this probe. You can ask students to draw a diagram to support their explanation. Students who choose Luna may describe a more detailed formation of the canyon that includes other factors that are part of the distracters, such as the role of plate tectonics, flash flooding, volcanic activity, wearing down of the mountains to form the plateau, and so on. However, the main factor that affected the formation of the Grand Canyon was the flow of river water.

Related Core Ideas in *Benchmarks for Science Literacy* (AAAS 2009)

3–5 Processes That Shape the Earth
- Waves, wind, water, and ice shape and reshape the Earth's land surface by eroding rock and soil in some areas and depositing them in other areas, sometimes in seasonal layers.

6–8 Processes That Shape the Earth
- Some changes in the Earth's surface are abrupt (such as earthquakes and volcanic eruptions) while other changes happen very slowly (such as uplift and wearing down of mountains).
- The Earth's surface is shaped in part by the motion of water (including ice) and wind over very long times, which acts to level mountain ranges. Rivers and glacial ice carry off soil and break down rock, eventually depositing the material in sediments or carrying it in solution to the sea.
- There are a variety of different land forms on the Earth's surface (such as coastlines, rivers, mountains, deltas, and canyons).

Related Core Ideas in *A Framework for K–12 Science Education* (NRC 2012)

K–2 ESS2.A: Earth Materials and Systems
- Wind and water can change the shape of the land.

3–5 ESS2.A: Earth Materials and Systems
- Rainfall helps to shape the land and affects the types of living things found in a region. Water, ice, wind, living organisms, and gravity break rocks, soils, and sediments into smaller particles and move them around.

6–8 ESS2.A: Earth Materials and Systems
- The planet's systems interact over scales that range from microscopic to global in size, and they operate over fractions of a second to billions of years. These interactions have shaped Earth's history and will determine its future.

6–8 ESS2.C: The Role of Water in Earth's Surface Processes
- Water's movements—both on the land and underground—cause weathering and erosion, which change the land's surface features and create underground formations.

Related Research

- Students of all ages may hold the view that Earth is the same now as when it was formed and that any changes must have been sudden and comprehensive (Freyberg 1985).
- Some students view Earth as static and unchanging (Cheek 2010).
- When interviewed about how a canyon was formed, some college students used catastrophic events in their explanation. They believed either a catastrophic event such as a flood, earthquake, or volcano formed a canyon or that a canyon started with a catastrophic event and then formed through river erosion. Some students also provided a biblical explanation of a giant flood (Sexton 2012).

- Some students think rivers get heavy and sink into Earth, thus carving out the land and the river (Mackintosh 2005).

Suggestions for Instruction and Assessment

- This probe can be extended by asking students to describe a model they could use to explain how the Grand Canyon was formed.
- Probe further by asking students how long they think it took for the canyon to form and what the area may have looked like millions of years ago.
- PBS Learning Media provides a good video on the formation of the Grand Canyon, including the role of the river system in eroding the canyon walls. The video is available at *www.pbslearningmedia. org/resource/ess05.sci.ess.earthsys.canyon/ the-grand-canyon-how-it-formed*.
- Because this probe is specific to the Grand Canyon, more questioning is needed to know if students would describe the same processes for other canyons in other settings. Show students pictures of other canyons throughout the world, and have

them explain how they think the canyons were formed.
- Have students research the role of wind in shaping some canyons.

References

American Association for the Advancement of Science (AAAS). 2009. Benchmarks for science literacy online. *www.project2061.org/publications/bsl/online*.

Cheek, K. 2010. Commentary: A summary and analysis of twenty-seven years of geoscience conceptions research. *Journal of Geoscience Education* 58 (3): 122–134.

Freyberg, P. 1985. Implications across the curriculum. In *Learning in science: The implications of children's science*, ed. R. Osborne and P. Freyberg, 125–135. Auckland, New Zealand: Heinemann.

Mackintosh, M. 2005. Children's understanding of rivers. *International Research in Geographical and Environmental Education* 14 (4): 316–322.

National Research Council (NRC). 2012. *A framework for K–12 science education: Practices, crosscutting concepts, and core ideas.* Washington, DC: National Academies Press.

Sexton, J. 2012. College students' conception of the role of rivers in canyon formation. *Journal of Geoscience Education* 60 (2): 168–178.

Mountains and Beaches

Four friends were walking on a beach. They picked up a handful of sand and looked at it closely. They noticed it was made mostly of very tiny grains of a white rock. One friend said, "My dad told me that when we walk on the sand, we are walking on the tops of mighty mountains. What do you think he meant by that?" This is what they said:

Christie: I think he meant this beach will become a tall mountain in 1,000 years.

Jonathan: I think he meant tiny pieces of rock from undersea volcanoes or mountains formed this beach.

Michael: I think he meant tiny pieces of rock from far away mountains on land ended up on the beach.

Liza: I think he meant there was once a mountain on this beach, but the mountain gradually eroded away and left the sand.

Who do you think has the best idea? _____ Explain your thinking.

_____

_____

_____

_____

Mountains and Beaches

Teacher Notes

Purpose

The purpose of this assessment probe is to elicit students' ideas about processes that change Earth. The probe is designed to find out if students recognize that the sand on many beaches results from weathering of rock on mountains, erosion, and deposition.

Type of Probe

Friendly talk

Related Concepts

Deposition, erosion, landforms, mountains, weathering

Explanation

The best answer is Michael's: "I think he meant that tiny pieces of rock from far away mountains on land ended up on the beach." There are many different types of sand of different colors that have different origins— rock or biogenic material, or both. The sand in this probe is composed of minerals from

rock. Landmasses, including mountains, are made up of rocks and minerals, such as quartz, feldspar, and mica. The most common component of most beach sand is silicon dioxide, in the form of quartz, which gives many beaches their white sand. Wind and rain constantly weather and erode mountain summits, hillsides, and other landforms that are made up of rocks. The fine grains of weathered material from mountains and inland rocks are transported by streams to the mouths of rivers, where they are carried into the ocean. Currents can carry these particles many miles, eventually depositing them along a shoreline. For example, although sand contains many materials, much of the sand on Florida's white, sandy beaches contains tiny quartz crystals that came from the weathering of the distant Appalachian Mountains. The quartz crystals are washed into and transported by rivers to the Atlantic Ocean and the Gulf of Mexico, where they are deposited onto the beaches by water currents and waves.

Administering the Probe

This probe is best used with grades 5–12. Sands around the world vary in composition and color, so be sure students understand that the sand in this example is made up of very tiny pieces of white rock. If you have a sample of beach sand containing white quartz, consider showing students the sand described in the probe. The probe can be extended by asking students to draw a model to support their explanation.

Related Core Ideas in *Benchmarks for Science Literacy* (AAAS 2009)

3–5 Processes That Shape the Earth

- Waves, wind, water, and ice shape and reshape the Earth's land surface by eroding rock and soils in some areas and depositing them in other areas, sometimes in seasonal layers.

6–8 Processes That Shape the Earth

- Some changes in the Earth's surface are abrupt (such as earthquakes and volcanic eruptions) while other changes happen very slowly (such as uplift and wearing down of mountains).

- The Earth's surface is shaped in part by the motion of water (including ice) and wind over very long times, which acts to level mountain ranges. Rivers and glacial ice carry off soil and break down rock, eventually depositing the material in sediments or carrying it in solution to the sea.

Related Core Ideas in *A Framework for K–12 Science Education* (NRC 2012)

3–5 ESS2.A: Earth Materials and Systems

- Rainfall helps to shape the land and affects the types of living things found in a region. Water, ice, wind, living organisms, and gravity break rocks, soils, and sediments into smaller particles and move them around.

- Earth's major systems are the geosphere (solid and molten rock, soil, and sediments), the hydrosphere (water and ice), the atmosphere (air), and the biosphere (living things, including humans). These systems interact in multiple ways to affect Earth's surface materials and processes. The ocean supports a variety of ecosystems and organisms, shapes landforms, and influences climate.

6–8 ESS2.C: The Role of Water in Earth's Surface Processes

- Water's movements—both on the land and underground—cause weathering and erosion, which change the land's surface features and create underground formations.

9–12 ESS2.C: The Role of Water in Earth's Surface Processes

- The abundance of liquid water on Earth's surface and its unique combination of physical and chemical properties are central to the planet's dynamics. These properties include water's exceptional capacity to absorb, store, and release large amounts of energy, transmit sunlight, expand upon freezing, dissolve and transport materials, and lower the viscosities and melting points of rocks.

Related Research

- The aspect of scale may be problematic for students. For example, comprehending the length of time it takes for mountains to erode is difficult for some students (Ault 1994).

- A study of college students' understanding of magnitudes of time associated with long-term geologic change, such as the formation of mountains, reveals difficulty comprehending temporal magnitudes (Lee et al. 2011).

- A study by Freyberg (1985) revealed that many students think Earth today is the same

as it has always been and that any changes to Earth (such as formation of a beach) were sudden and comprehensive. However, it is important to note that students in this study did not have formal instruction in the topics addressed (AAAS 1993).

- Happs (1982) found students tend to use different meanings for rock fragments than scientists, who classify the fragments by average size. For example, instead of using particle size to distinguish between boulders, gravel, sand, and clay, students associate the particles with their origin. Sand is defined as coming from a beach or desert rather than being a particle of a certain average size (Driver et al. 1994).

Suggestions for Instruction and Assessment

- Combine this probe with "Beach Sand" from *Uncovering Student Ideas in Science, Volume 1* (Keeley, Eberle, and Farrin 2005).
- Have students define sand and distinguish it from other particles of rock origin.
- Have students examine samples of sand from different areas and containing different mineral and biogenic material. Connect the materials the sand is made up of to their origin, and develop ideas about how the sand ended up forming a beach. Elementary students can examine sand with magnifiers to observe the particles and compare the pieces to actual rock samples, particularly ones that contain recognizable minerals, such as quartz or mica. This helps them understand the origin of sand from rock, and the rock can be later traced back to landforms such as mountains and exposed bedrock.
- Combine students' observations of tumbling solid rock or of eroding sand in a stream table with the role these phenomena play in shaping the surface of Earth (e.g., beaches, wearing down of mountains,

widening of rivers). When using stream tables to investigate erosion and deposition phenomena, be sure to trace back to the origin of the sand. It is also important to explicitly address the fact that the processes the stream tables are modeling in a very short time actually occur over long periods of time.

- Combine an understanding of the use of models with an understanding of the process of beach formation. This is a good time for students to understand how models are used, as well as their limitations in representing phenomena.
- Provide students with a single grain of sand from a local (or the closest) beach and ask them to trace back to the origin of the grain of sand. What journey do they think it took before ending up on the beach?

References

American Association for the Advancement of Science (AAAS). 1993. *Benchmarks for science literacy.* New York: Oxford University Press.

American Association for the Advancement of Science (AAAS). 2009. Benchmarks for science literacy online. *www.project2061.org/publications/bsl/online.*

Ault, C. R. 1994. Research in problem solving in Earth science. In *Handbook of research on science teaching and learning,* ed. D. Gabel, 269–283. New York: Simon and Schuster.

Driver, R., A. Squires, R. Rushworth, and V. Wood-Robinson. 1994. *Making sense of secondary science: Research into children's ideas.* London and New York: RoutledgeFalmer.

Freyberg, P. 1985. Implications across the curriculum. In *Learning in science: The implications of children's science,* ed. R. Osborne and P. Freyberg, 125–135. Auckland, New Zealand: Heinemann.

Happs, J. 1982. Some aspects of student understanding of rocks and minerals. Science Education Research Unit Working Paper 204, University of Waikato, Hamilton, New Zealand.

Keeley, P., F. Eberle, and L. Farrin. 2005. *Uncovering student ideas in science, vol. 1: 25 formative assessment probes.* Arlington, VA: NSTA Press.

Lee, H., O. Liu, C. Price, and A. Kendall. 2011. College students' temporal magnitude recognition ability associated with durations of scientific changes. *Journal of Research in Science Teaching* 48 (3): 317–335.

National Research Council (NRC). 2012. *A framework for K–12 science education: Practices, crosscutting concepts, and core ideas.* Washington, DC: National Academies Press.

How Do Rivers Form?

Five friends were fishing along a river. They wondered how rivers formed. They each had different ideas. This is what they said:

Herb: I think most rivers formed when Earth was formed. They have always been there. They have just gotten bigger over time.

Jack: I think most rivers formed from a path or cracks in the ground. The water fills in the path or cracks and because of its weight, sinks down into the ground and creates a river.

Ricardo: I think most rivers formed when ocean water collected in a tiny pool. The pool gets larger and creates a small stream that flows toward land. Fresh water gets added to the stream. Over time, this stream gets wider and forms a river.

Kasper: I think most rivers formed when water collected and formed small streams. These small streams collect more water, get bigger, meet each other, and form bigger streams. Eventually, the bigger streams meet to form rivers.

Cooper: I think most rivers formed from huge floods or glaciers that happened millions of years ago. The floods or glaciers carved out the land by carrying away rocks and soil, and the water filled it in.

Which friend do you agree with the most? _____ Explain why you agree.

_____

_____

_____

_____

How Do Rivers Form?

Teacher Notes

Purpose

The purpose of this assessment probe is to elicit students' ideas about rivers. The probe is designed to find out how students think most rivers formed.

Type of Probe

Friendly talk

Related Concepts

Erosion, river systems, tributaries

Explanation

The best answer is Kasper's: "I think most rivers formed when water collected and formed small streams. These small streams collect more water, get bigger, meet each other, and form bigger streams. Eventually, the bigger streams meet to form rivers." Rivers come in different shapes and sizes, but the one thing they all have in common is they start at a higher elevation, such as a mountain, hill, or some other surface above sea level. The beginning source of water can come from snow and ice melt, a spring, or even a high-elevation lake such as Lake Itasca in Minnesota from which the Mississippi River starts.

The river water starts at a high point and flows downward toward lower elevations in a small stream. As it flows downward, it picks up more water from rain and other small streams that join, thus forming a larger stream. These streams continue to join with other streams forming larger bodies, called *tributaries*, as more water is added. These tributaries join to form the major rivers. As streams and rivers flow, they cut into the land through the process of erosion. Over time, rivers change the land as they carve out new or wider paths.

Administering the Probe

This probe can be used with grades 3–12. The probe intentionally avoids use of terminology such as *tributaries* and *erosion* to focus on students' conceptual understanding. It is helpful to show a photograph of a river when

presenting the probe. A local river can be used as the context for this probe.

Related Core Ideas in *Benchmarks for Science Literacy* (AAAS 2009)

3–5 Processes That Shape the Earth

- Waves, wind, water, and ice shape and reshape the Earth's land surface by eroding rock and soil in some areas and depositing them in other areas, sometimes in seasonal layers.

6–8 Processes That Shape the Earth

- The Earth's surface is shaped in part by the motion of water (including ice) and wind over very long times, which acts to level mountain ranges. Rivers and glacial ice carry off soil and break down rock, eventually depositing the material in sediments or carrying it in solution to the sea.
- There are a variety of different land forms on the Earth's surface (such as coastlines, rivers, mountains, deltas, and canyons).

Related Core Ideas in *A Framework for K–12 Science Education* (NRC 2012)

3–5 ESS2.A: Earth Materials and Systems

- Rainfall helps to shape the land and affects the types of living things found in a region. Water, ice, wind, living organisms, and gravity break rocks, soils, and sediments into smaller particles and move them around.

6–8 ESS2.C: The Roles of Water in Earth's Surface Processes

- Water continually cycles among land, ocean, and atmosphere via transpiration, evaporation, condensation and crystallization, and precipitation, as well as downhill flows on land.

- Water's movements—both on the land and underground—cause weathering and erosion, which change the land's surface features and create underground formations.

Related Research

- Students of all ages may hold the view that Earth is the same now as it was when it was formed and that any changes that have occurred must have been sudden and comprehensive (Freyberg 1985).
- Some students view Earth as static and unchanging (Cheek 2010).
- Some students may think that a river fills out a previously existing path (Martinez, Bannan, and Kitsantas 2012).
- Some students think rivers get heavy and sink into Earth, thus carving out the land and the river (Mackintosh 2005).

Suggestions for Instruction and Assessment

- Have students search for and examine maps of major rivers and their tributaries to see how smaller rivers and streams flow and merge into larger streams and rivers.
- The clarification statement for the *Next Generation Science Standards* fourth-grade performance expectation ESS2-1, "Make observations and/or measurements to provide evidence of the effects of weathering or the rate of erosion by water, ice, wind, or vegetation," suggests that one of the variables students can test is the angle of slope in the downhill movement of water (NGSS Lead States 2013).
- Ask students how an interstate highway can be used as an analogy (conceptual model) for a river system. How does the model explain a river system? What are the limitations of this model?
- Use stream tables to model river formation, water flow, and erosion. Simple stream

tables can be made with aluminum foil pans and sand.

- Be aware that some students may have had early childhood experience through play that is similar to river formation, such as playing "dams and rivers" on sloping dirt driveways after it rains or pouring water on sand structures in sandboxes or beach play. Use those experiences as springboards for thinking about how rivers form and their role in erosion and deposition.

- River Cutters is a middle school GEMS curriculum guide that contains several inquiry-based lessons on how rivers form, including the factors of slope and amount of precipitation. For a review of this curriculum guide, see NSTA Recommends at *www.nsta.org/recommends/ViewProductPrint. aspx?ProductID=11936*.

References

American Association for the Advancement of Science (AAAS). 2009. Benchmarks for science literacy online. *www.project2061.org/publications/bsl/online.*

Cheek, K. 2010. Commentary: A summary and analysis of twenty-seven years of geoscience conceptions research. *Journal of Geoscience Education* 58 (3): 122–134.

Freyberg, P. 1985. Implications across the curriculum. In *Learning in science: The implications of children's science,* ed. R. Osborne and P. Freyberg, 125–135. Auckland, New Zealand: Heinemann.

Mackintosh, M. 2005. Children's understanding of rivers. *International Research in Geographical and Environmental Education* 14 (4): 316–322.

Martinez, P., B. Bannan, and A. Kitsantas. 2012. Bilingual students' ideas and conceptual change about slow geomorphological changes caused by water. *Journal of Geoscience Education* 60 (1): 54–67.

National Research Council (NRC). 2012. *A framework for K–12 science education: Practices, crosscutting concepts, and core ideas.* Washington, DC: National Academies Press.

NGSS Lead States. 2013. *Next Generation Science Standards: For states, by states.* Washington, DC: National Academies Press. *www.nextgenscience. org/next-generation-science-standards.*

What Is the Inside of Earth Like?

Five friends were talking about Earth's interior. They each had different ideas about what the inside of Earth was like. This is what they said:

Rudy: I think we would see layers. Most of the inside of Earth will be hot liquid.

Lizette: I think we would see layers. Most of the inside of Earth will be solid.

Zara: I think we would see three layers with a giant magnet in the center of Earth.

Mateo: I don't think there are layers. Earth is made up of rocks and dirt with hot liquid found in the cracks.

Haliaka: I think we would see sections of solid and liquid Earth with gaps in between.

Who do you think has the best idea? _____ Explain your thinking. You may draw a picture to show what you think the inside of Earth is like.

_____

_____

_____

_____

What Is the Inside of Earth Like?

Teacher Notes

Purpose

The purpose of this assessment probe is to elicit students' ideas about Earth's structure. The probe is designed to find out if students recognize Earth's interior is made up of layers and is mostly solid.

Type of Probe

Friendly talk

Related Concepts

Earth's crust, inner and outer core, mantle, structure of Earth

Explanation

The best answer is Lizette's: "I think we would see layers. Most of the inside of Earth will be solid." Realistically, we cannot see all the layers of Earth. The deepest hole ever drilled was about 12 km deep. However, scientists are able to use indirect methods such as seismic waves to determine the composition of the deep interior of Earth. Starting from the center, Earth is composed of these layers: inner core, outer core, mantle, and crust. The crust is mostly made up of solid, rigid rock with a very thin layer of soil. The mantle, where temperatures are much greater than in the crust, is mostly made up of solid and semi-solid rock. Although semi-solid rock is still considered solid, it is "softened" and deformable. The outer core is the only liquid layer (not to be confused with magma), and it is mainly made up of molten iron and nickel. The inner core, which is under tremendous pressure, is believed to be a solid that is composed of an iron-nickel alloy.

Administering the Probe

This probe is best used with grades 6–12. Ask students to imagine what the inside of Earth would look like if you were able to slice Earth in half. Explain what a cross-sectional diagram is, and encourage students to draw and label a cross-sectional diagram of Earth from the center of Earth to where they are standing to support their explanation.

Related Core Ideas in *Benchmarks for Science Literacy* (AAAS 2009)

6–8 Processes That Shape the Earth
- The Earth first formed in a molten state and then the surface cooled into solid rock.

6–8 The Earth
- The Earth is mostly rock.

Related Core Ideas in *A Framework for K–12 Science Education* (NRC 2012)

9–12 ESS2.A: Earth Materials and Systems
- Evidence from deep probes and seismic waves, reconstructions of historical changes in Earth's surface and its magnetic field, and an understanding of physical and chemical processes lead to a model of Earth with a hot but solid inner core, a liquid outer core, a solid mantle and crust. Motions of the mantle and its plates occur primarily through thermal convection, which involves the cycling of matter due to the outward flow of energy from Earth's interior and gravitational movement of denser materials toward the interior.

Related Research

- One common idea about Earth's structure is that Earth is mostly molten rock, including the core of Earth. Some students believe this is where magma comes from. Elementary and middle school science curricula frequently introduce students to the different layers of Earth. However, research shows that many students, after being taught about Earth's structure, leave with misconceptions about where the layers are and the composition of the layers (Sharpe, Mackintoch, and Seedhouse 1995).
- Some students think Earth's crust is much thicker than what it really is. When asked to draw the layers of Earth, most students drew concentric circles, but their drawings showed little understanding of the proportional size of the layers. Some students showed lack of a continuous model of the inner structure of Earth, thinking there are "hollow spaces" or gaps between layers. Some students even showed a magnet in the center of Earth (Lillo 1994).
- A sample of 120 Turkish students ages 13–14 revealed alternative conceptions about the structure of Earth. In most cases, students thought (1) Earth is composed of three layers: the crust, the core, and an in-between zone; (2) Earth is filled with either water or magma; and (3) the core serves as the source of magma (Dal 2007).
- Undergraduate students in an introductory geology course were asked to draw a picture of Earth's interior and provide think-aloud explanations of their drawings. The results revealed that students hold a wide range of alternative conceptions about Earth, with only a small fraction having scientifically accurate ideas. Students' understandings ranged from conceptualizing Earth's interior as consisting of horizontal layers of rock and dirt, to more sophisticated views such as Earth's interior being composed of concentric layers with unique physical and chemical characteristics. Of the students studied, 38% thought the core was all liquid (McAllister 2014).

Suggestions for Instruction and Assessment

- Use the annotated drawing strategy to check for understanding after students have had an opportunity to learn about the structure of Earth (Keeley 2015). Ask students to imagine they could cut Earth in half with a knife. Have them draw what they would expect the inside to look like and then label and describe the layers.

- Discuss how popular media may have led to the idea that Earth is mostly molten rock. Movies and fictional stories that include "journeys to the center of Earth" often portray dramatic rises in temperature, fire inside Earth, and molten rock.

- Developing an understanding of how volcanoes develop and where magma comes from may help dispel the idea that there are only two layers—a solid outer layer and an inner molten layer of rock. Students' images of erupting volcanoes and red hot lava, combined with their experience in seeing only the very surface of Earth, may contribute to this misconception of two layers—one solid, one liquid.

- Have students research and describe how scientists learn about the inner structure of Earth.

- Some older students might know that Earth has a magnetic field and magnetic poles, leading them to believe that a giant magnet is inside Earth. Have students research websites to find out how the composition and properties of the core are related to Earth's magnetic field. The Natural Resources of Canada website at *www.geomag.nrcan.gc.ca/mag_fld/fld-en.php* and the U.S. Geological Survey Geomagnetism Program site at *http://geomag.usgs. gov* provide information to help students understand how Earth acts like a magnet.

References

American Association for the Advancement of Science (AAAS). 2009. Benchmarks for science literacy online. *www.project2061.org/publications/bsl/online*.

Dal, B. 2007. How do we help students build beliefs that allow them to avoid critical learning barriers and develop a deep understanding of geology? *Eurasia Journal of Mathematics Science and Technology Education* 3 (4): 251–269.

Keeley, P. 2015. *Science formative assessment: 75 practical strategies for linking assessment, instruction, and learning*. 2nd ed. Thousand Oaks, CA: Corwin Press.

Lillo, J. 1994. An analysis of the annotated drawings of the internal structure of the Earth made by students ages 10 to 15 from primary and secondary schools in Spain. *Teaching Earth Sciences* 19 (3): 83–87.

McAllister, M. 2014. A study of undergraduate students' alternative conceptions of Earth's interior using drawing tasks. *Journal of Astronomy and Earth Sciences Education* 1 (1): 23–36.

National Research Council (NRC). 2012. *A framework for K–12 science education: Practices, crosscutting concepts, and core ideas*. Washington, DC: National Academies Press.

Sharpe, J., M. Mackintoch, and P. Seedhouse. 1995. Some comments on children's ideas about Earth structure, volcanoes, earthquakes, and plates. *Teaching Earth Sciences* 20 (1): 28–30.

Describing Earth's Plates

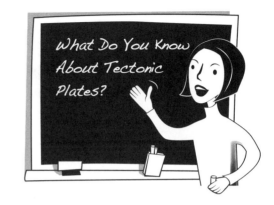

What Do You Know About Tectonic Plates?

What do you know about Earth's tectonic plates? Put an X in front of each statement that describes Earth's tectonic plates.

_____ **A.** Plates are huge, rigid slabs of rock.

_____ **B.** Plates are different sizes.

_____ **C.** Plates are all about the same thickness.

_____ **D.** Plate boundaries form most of the coastlines of the world.

_____ **E.** Plates move very slowly.

_____ **F.** Plates float on top of hot liquid magma.

_____ **G.** Some of the plates have large gaps between them.

_____ **H.** Plates can include both continental crust and ocean basin.

_____ **I.** Plates are stacked in layers.

_____ **J.** Each plate forms a continent.

_____ **K.** Plates cover most of Earth but not all of it.

_____ **L.** Patterns of earthquakes and volcanoes can be found along plate boundaries.

_____ **M.** Plate boundaries can be found in the middle of the ocean.

_____ **N.** Plates are made up of hot material.

_____ **O.** Plates are found deep within Earth.

_____ **P.** Mountain chains formed along plate boundaries.

Explain your thinking. Describe what you know about Earth's plates.

_____

_____

_____

_____

Describing Earth's Plates

Teacher Notes

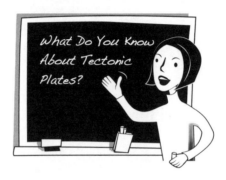

What Do You Know About Tectonic Plates?

Purpose

The purpose of this assessment probe is to elicit students' ideas about Earth's tectonic plates. The probe is designed to uncover commonly held misconceptions about Earth's plates.

Type of Probe

Justified list

Related Concepts

Earthquakes, plate tectonics, structure of Earth, volcanoes

Explanation

The best answers are A, B, E, H, L, M, and P. All of Earth is covered with huge, rigid slabs or fragments of rock called tectonic plates. The outer portion of Earth's lithosphere, including the continental crust and the oceanic crust that composes ocean basins, and the upper mantle make up these huge plates of solid rock. They can be different sizes. For example, the North American Plate is quite large, and the Juan de Fuca Plate is one of the smallest plates. Thickness also varies. Oceanic crust tends to be thinner than continental crust. Some boundaries appear along coastlines, but other boundaries can be in the ocean far away from land. The plates do not form the continents but they do move them, and some plates, such as the Nazca Plate, are made only of oceanic crust and do not include a continental land mass. Plates move slowly. Some move at about the same rate that your fingernails grow (2.5 cm per year). Others, such as at the East Pacific Rise, move at more than 15 cm per year. Plates do not float on top of hot, liquid magma, but they do move above a hot, deformable layer of "soft" rock that is almost like a solid that flows. Interaction between plates often results in events and features such as earthquakes and volcanoes. These can sometimes appear in a pattern forming plate boundaries, such as the Pacific Ring of Fire. However, there are also intra-plate earthquakes and volcanoes that appear at hot spots that are not on plate

boundaries. Mountain chains form along plate boundaries. However, although mountain chains form along plate boundaries, they may no longer be at an active plate boundary. The Appalachian Mountains are an example.

Administering the Probe

This probe is best used with grades 5–12. Encourage students to explain their evidence and reasoning for the statements they selected, as well as their evidence and reasoning for not choosing some of the statements. This probe can be combined with the card sort formative assessment classroom technique described in the classroom snapshot on pages 3–4.

Related Core Ideas in *Benchmarks for Science Literacy* (AAAS 2009)

6–8 Processes That Shape the Earth

- The outer layer of the Earth—including both the continents and the ocean basins—consists of separate plates.
- The Earth's plates sit on a dense, hot, somewhat melted layer of the Earth. The plates move very slowly, pressing against one another in some places and pulling apart in other places, sometimes scraping alongside each other as they do.
- There are worldwide patterns to major geological events (such as earthquakes, volcanic eruptions, and mountain building) that coincide with plate boundaries.

9–12 Processes That Shape the Earth

- The outward transfer of the Earth's internal heat causes regions of different temperatures and densities. The action of a gravitational force on regions of different densities causes the rise and fall of material between the Earth's surface and interior, which is responsible for the movement of plates.
- Earthquakes often occur along the boundaries between colliding plates, and molten rock

from below creates pressure that is released by volcanic eruptions, helping to build up mountains. Under the ocean basins, molten rock may well up between separating plates to create new ocean floor. Volcanic activity along the ocean floor may form undersea mountains, which can thrust above the ocean's surface to become islands.

Related Core Ideas in *A Framework for K–12 Science Education* (NRC 2012)

3–5 ESS2.B: Plate Tectonics and Large-Scale System Interactions

- The locations of mountain ranges, deep ocean trenches, ocean floor structures, earthquakes, and volcanoes occur in patterns. Most earthquakes and volcanoes occur in bands that are often along the boundaries between continents and oceans. Major mountain chains form inside continents or near their edges. Maps can help locate the different land and water features areas of Earth.

6–8 ESS2.B: Plate Tectonics and Large-Scale System Interactions

- Maps of ancient land and water patterns, based on investigations of rocks and fossils, make clear how Earth's plates have moved great distances, collided, and spread apart.

9–12 ESS2.B: Plate Tectonics and Large-Scale System Interactions

- Plate tectonics is the unifying theory that explains the past and current movements of the rocks at Earth's surface and provides a framework for understanding its geologic history. Plate movements are responsible for most continental and ocean-floor features and for the distribution of most rocks and minerals within Earth's crust.

Related Research

- In a study of Portuguese students ages 16–17, many students thought that coastlines or continental boundaries are the boundaries between tectonic plates. They also thought that continents float on top of hot, liquid material (Marques and Thompson 1997).

- As part of an assessment project, Horizon Research Inc. (HRI) interviewed middle school students about their plate tectonics ideas. Several misconceptions surfaced, including (1) large gaps or spaces exist between plates; (2) plates stack like dinner plates; (3) a continent is a plate, and plates are the same shape as continents; and (4) continents do not move. These students did recognize that earthquakes, volcanoes, and mountains seem to form in the same area, but they had no explanation for this (Ford and Taylor 2006).

- College students enrolled in introductory geosciences courses were unsure about the location of tectonic plates, describing them as existing below Earth's surface, at the core, or even in the atmosphere (Libarkin et al. 2005).

- Several plate misconceptions are listed, along with performance data on assessment questions, on the American Association for the Advancement of Science (AAAS) assessment website at *http://assessment.aaas. org/topics/PT#/,tabs-204/2.*

Suggestions for Instruction and Assessment

- Ask students to draw a side-view picture of what they think a plate looks like. This picture may give you additional insight into students' thinking.

- Provide students with a map of Earth's plates. Have them revisit their answer choices after observing the boundaries of the plates. Engage them in a discussion about what they see on these maps.

- Help students envision a plate as a single slab of continental or oceanic crust, rather than a floating slab of "land."

- If students believe that the continents are plates, refer them to the map of Earth's plates. Help them see that the plates cover all of Earth, not just the continents.

- Have students identify (1) plates that include continents and ocean basins and (2) plates that include only ocean basins. Have them identify plate boundaries that are in the middle of the ocean. Ask them if they can find any plates that are made only of continental crust.

- Have students plot earthquake and volcanic activity on a map that shows plate boundaries. Have them describe the patterns they observe.

- Compare and contrast the sizes of different plates.

- Search the Internet for plate tectonics interactive maps, simulations, and videos that can be used to confront students with their initial ideas about tectonic plates. For example, PBS Learning Media has several educator resources at *www.pbslearningmedia.org/resource/ess05.sci.ess.earthsys. lp_platetectonics/plate-tectonics.*

- The U.S. Geological Survey's comprehensive book on plate tectonics, *This Dynamic Earth: The Story of Plate Tectonics,* can be accessed online at *http://pubs.usgs.gov/gip/dynamic.*

- It is recommended that representations used to teach plate tectonics, such as diagrams in textbooks, be carefully analyzed and revised in light of researchers' findings. In particular, some aspects of plate tectonic images, such as the fading slab tips and orange mantle, may induce or reinforce misconceptions (Clark et al. 2011).

- Additional assessment items on Earth's plates, along with listings of sub-ideas and performance data can be accessed from AAAS Project 2061 at *http://assessment. aaas.org/topics/PT#/,tabs-204/1.*

- For reviewed phenomena and representations for plate tectonics, go to the PRISMS

(Phenomena and Representations for the Instruction of Science in Middle Schools) website at *http://prisms.mmsa.org/browse.php?cat=2*.

- The AAAS Plate Tectonics Strand Map is useful for examining the progression of ideas and the connections between them. This map can be accessed at *http://strandmaps.dls.ucar.edu/?id=SMS-MAP-0049*.

References

American Association for the Advancement of Science (AAAS). 2009. Benchmarks for science literacy online. *www.project2061.org/publications/bsl/online*.

Clark, S., J. Libarkin, K. Kortz, and S. Jordan. 2011. Alternative conceptions of plate tectonics held by nonscience undergraduates. *Journal of Geoscience Education* 59 (4): 251–262.

Ford, B., and M. Taylor. 2006. Investigating students' ideas about plate tectonics. *Science Scope* 30 (1): 38–43.

Libarkin, J., S. Anderson, J. Dahl, M. Beilfuss, W. Boone, and J. Kurdziel. 2005. College students' ideas about geologic time, Earth's interior, and Earth's crust. *Journal of Geoscience Education* 53 (1): 17–26.

Marques, L., and D. Thompson. 1997. Misconceptions and conceptual changes concerning continental drift and plate tectonics among Portuguese students aged 16–17. *Research in Science and Technological Education* 15 (2): 195–222.

National Research Council (NRC). 2012. *A framework for K–12 science education: Practices, crosscutting concepts, and core ideas*. Washington, DC: National Academies Press.

Where Do You Find Earth's Plates?

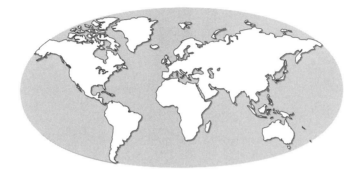

Six students were learning about plate tectonics in their science class. They looked at a map of the world and wondered if geography determined where Earth's tectonic plates were found. Each student had different ideas about the location of Earth's plates. This is what they said:

Ari: I think plates include continents, but not ocean floor.

Sergio: I think plates include ocean floor, but not continents.

Miriam: I think all plates include both continents and ocean floor.

Akoni: I think some plates include both continents and ocean floor, and other plates include only ocean floor. There are no plates that include only continents.

Lila: I think some plates include only continents; some include only ocean floor, and others include both continents and ocean floor.

Ursula: I think none of the plates include continents or the ocean floor. They are found somewhere else.

Who do you think has the best idea? _____ Explain your thinking.

_____

_____

_____

_____

Where Do You Find Earth's Plates?

Teacher Notes

Purpose

The purpose of this assessment probe is to elicit students' ideas about Earth's plates. The probe is designed to find out where students think tectonic plates are located.

Type of Probe

Friendly talk

Related Concepts

Continental crust, ocean basin crust, plate tectonics

Explanation

The best answer is Akoni's: "I think some plates include both continents and ocean floor, and other plates include only ocean floor. There are no plates that include only continents." No plates are made of only continental crust. Earth is composed of several major, large tectonic plates and several smaller ones. These plates are made up of rigid rock from the lithosphere, which consists of Earth's crust and the uppermost mantle. When you look at a map of Earth with the plate boundaries drawn in, you can see that some plates include both a continent (or part of a continent) and an ocean basin. For example, the North American Plate, the Pacific Plate, and the African Plate include both continental crust and ocean basin crust. Other plates include only ocean basin crust. Examples of plates composed of ocean basin crust that do not include continental crust are the Nazca Plate and the Philippines Plate.

Administering the Probe

This probe is best used with grades 6–12. Make sure that students understand that the probe is asking where the plates are found.

Related Core Ideas in *Benchmarks for Science Literacy* (AAAS 2009)

6–8 Processes That Shape the Earth
- The outer layer of the Earth—including both the continents and the ocean basins—consists of separate plates.

Related Core Ideas in *A Framework for K–12 Science Education* (NRC 2012)

3–5 ESS2.B: Plate Tectonics and Large-Scale System Interactions
- The locations of mountain ranges, deep ocean trenches, ocean floor structures, earthquakes, and volcanoes occur in patterns. Most earthquakes and volcanoes occur in bands that are often along the boundaries between continents and oceans. Major mountain chains form inside continents or near their edges. Maps can help locate the different land and water features areas of Earth.

6–8 ESS1.C: The History of Planet Earth
- Tectonic processes continually generate new ocean sea floor at ridges and destroy old sea floor at trenches.

9–12 ESS2.B: Plate Tectonics and Large-Scale System Interactions
- Plate tectonics is the unifying theory that explains the past and current movements of the rocks at Earth's surface and provides a framework for understanding its geologic history. Plate movements are responsible for most continental and ocean-floor features and for the distribution of most rocks and minerals within Earth's crust.

Related Research

- When children see a world map, they typically focus on the continental land masses and do not consider that the ocean floor is also made up of Earth's crust. As a result, many students consider the continents as being the plates (Sharpe et al. 1995).
- A study by Horizon Research Inc. (HRI) interviewed students about their ideas related to plate tectonics. Several misconceptions surfaced, including (1) a continent is a plate, (2) plates are the same shape as the continents, (3) continents have no relation to plates, (4) plates are under the ocean but do not include the continents, and (5) plates are somehow "down there" and not really related to Earth's surface (Ford and Taylor 2006).

Suggestions for Instruction and Assessment

- Provide students with a map of the world's plates, such as the one provided by the U.S. Geological Survey at *http://pubs.usgs.gov/gip/dynamic/slabs.html*. Have students examine continental borders and areas covered by ocean basins and then identify what makes up each plate.
- Have students identify plate boundaries that occur in the middle of the ocean, such as the Pacific Plate.
- Have students identify continental coastlines that coincide with plate boundaries and continental coastlines that do not coincide with plate boundaries.
- Be aware that students' experiences consist primarily of looking at world maps that show continents, the ocean, and other bodies of water. Many students have no conception of what the ocean floor looks like. Have students describe what they think they would see if all the water was removed from the ocean. Have them draw and describe what they think the bottom of the ocean looks like. Then, show them maps of the ocean floor.

- Using a video from the Scripps Institution of Oceanography, students can see what the ocean basins would look like if all the water were drained away. Engage students in discussion about how mountain chains, volcanoes, and deep trenches formed in ocean basins, thus connecting those formations to plate tectonics. The video is available at *www.youtube.com/watch?v=MSbIKp571Rk.*

- Students may not understand that Earth's crust or its lithosphere covers all of Earth. They may think it covers only land masses and not areas covered by ocean. Connect this probe to understanding the structure of Earth.

References

American Association for the Advancement of Science (AAAS). 2009. Benchmarks for science literacy online. *www.project2061.org/publications/bsl/online.*

Ford, B., and M. Taylor. 2006. Investigating students' ideas about plate tectonics. *Science Scope* 30 (1): 38–43.

National Research Council (NRC). 2012. *A framework for K–12 science education: Practices, crosscutting concepts, and core ideas.* Washington, DC: National Academies Press.

Sharpe, J., M. Mackintoch, and P. Seedhouse. 1995. Some comments on children's ideas about Earth structure, volcanoes, earthquakes, and plates. *Teaching Earth Sciences* 20 (1): 28–30.

What Do You Know About Volcanoes and Earthquakes?

Erupting volcanoes and destructive earthquakes make newspaper headlines. What do you know about volcanoes and earthquakes? Put an X next to statements that describe volcanoes or earthquakes.

_____ **A.** Earthquakes occur more often in some places than others.

_____ **B.** Volcanoes occur in warm climates, earthquakes occur in both cold and warm climates.

_____ **C.** Volcanic eruptions are usually followed by earthquakes.

_____ **D.** Earthquakes cause most volcanic eruptions.

_____ **E.** Earthquakes and volcanoes show similar patterns in their locations.

_____ **F.** Earthquakes and volcanoes always occur at or near tectonic plate boundaries.

_____ **G.** Earthquakes are caused when tectonic plates crash into each other. The bigger the crash; the bigger the earthquake.

_____ **H.** Earthquakes eventually cause continents to break up and form plates.

_____ **I.** Earthquakes and volcanoes can occur in the ocean far from land.

_____ **J.** Earthquakes and volcanoes are found on every continent, including Antarctica.

_____ **K.** Earthquakes occur during earthquake weather.

_____ **L.** All mountains were once volcanoes.

Explain your thinking. Describe what you know about earthquakes and volcanoes.

_____

_____

_____

What Do You Know About Volcanoes and Earthquakes?

Teacher Notes

Purpose

The purpose of this assessment probe is to elicit students' ideas about plate tectonic features and events. The probe is designed to uncover commonly held misconceptions about volcanoes and earthquakes.

Type of Probe

Justified list

Related Concepts

Earthquakes, plate tectonics, volcanoes

Explanation

The best answers are A, E, I, and J. Earthquakes and volcanoes can occur almost anywhere on Earth, in both warm and cold environments (including Antarctica), but they tend to occur more frequently in certain areas and in a pattern along the boundaries of tectonic plates. For example, if you were to examine a map of the boundaries of the plates and superimpose volcanic and earthquake activity on the same

map, you would see that many volcanoes and earthquakes tend to fall right on or along those boundaries. One striking example is the string of volcanoes and seismic activity along the Pacific Plate. This string is called the Ring of Fire. About 90% of all earthquakes and 75% of all volcanic activity on Earth occur there. However, some earthquakes and volcanoes are located on continents or ocean floor far away from plate boundaries. A map of earthquake activity shows earthquake activity in the middle of the ocean. For example, the devastating tsunami that occurred in 2004 and slammed into 11 countries from Thailand to the African continent was the result of a powerful earthquake in the Indian Ocean. Volcanoes can also occur far from land. Many active and extinct volcanoes are under water on the ocean floor.

That earthquake activity has a causal effect on volcanic eruptions or vice versa is a common misconception, particularly the idea that one *usually* causes the other. Both are dramatic and often catastrophic phenomena, but neither is

usually caused by the other. However, some earthquakes are volcanically triggered, and typically those earthquakes are smaller than earthquakes not caused by volcanic sources.

Earthquakes are not the result of plates rapidly crashing into each other, and they do not break up continents to form plates. They occur because of the friction generated when plates rub against each other, but it is not a violent crash. Sometimes rock is snapped off in this process and enormous amounts of energy are released and transmitted as seismic waves that are felt as earthquakes. Volcanic eruptions are caused by a different process. Deep within Earth, some rock slowly melts and becomes magma. This magma contains gases. As magma rises and collects, it causes pressure on the surrounding rock and finds a way to the surface, where it is released as lava (as well as ash, rocks, and gases). One event does not cause the other, yet they are related because both often occur along plate boundaries and sometimes occur close to each other. Mountain formation is also related to tectonic processes. Some mountains are active, dormant, or extinct volcanoes, but mountains are also formed in other ways. Sometimes tectonic plates press against each other causing an uplift, which results in mountain formation. Earthquakes are not forecast like weather; there is no such thing as "earthquake weather."

Administering the Probe

This probe is best used with grades 3–12. This probe can be combined with the card sort or claim cards formative assessment classroom techniques (see pp. 3–4).

Related Core Ideas in *Benchmarks for Science Literacy* (AAAS 2009)

6–8 Processes That Shape the Earth
- The interior of the Earth is hot. Heat flow and movement of material within the Earth

cause earthquakes and volcanic eruptions and create mountains and ocean basins. Gas and dust from large volcanoes can change the atmosphere.
- Some changes in the Earth's surface are abrupt (such as earthquakes and volcanic eruptions) while other changes happen very slowly (such as uplift and wearing down of mountains).
- There are worldwide patterns to major geological events (such as earthquakes, volcanic eruptions, and mountain building) that coincide with plate boundaries.

9–12 Processes That Shape the Earth
- Earthquakes often occur along the boundaries between colliding plates, and molten rock from below creates pressure that is released by volcanic eruptions, helping to build up mountains. Under the ocean basins, molten rock may well up between separating plates to create new ocean floor. Volcanic activity along the ocean floor may form undersea mountains, which can thrust above the ocean's surface to become islands.

Related Core Ideas in *A Framework for K–12 Science Education* (NRC 2012)

3–5 ESS2.B: Plate Tectonics and Large-Scale System Interactions
- The locations of mountain ranges, deep ocean trenches, ocean floor structures, earthquakes, and volcanoes occur in patterns. Most earthquakes and volcanoes occur in bands that are often along the boundaries between continents and oceans. Major mountain chains form inside continents or near their edges. Maps can help locate the different land and water features areas of Earth.

9–12 ESS2.B: Plate Tectonics and Large-Scale System Interactions
- Plate tectonics is the unifying theory that explains the past and current movements of

the rocks at Earth's surface and provides a framework for understanding its geologic history. Plate movements are responsible for most continental and ocean-floor features and for the distribution of most rocks and minerals within Earth's crust.

Related Research

- Some students think that earthquakes shake the region around volcanoes and cause them to erupt. Researchers also found that some students think all volcanoes are on top of fault lines or weak spots on Earth and that "stuff just comes up there," and that all mountains were once volcanoes (Driver et al. 1994).

- Some students believe that earthquakes occur because of high winds or tornadoes (Allen 2010).

- The idea that volcanoes occur only in warm areas of the world is a common misconception held by both students and adults (Dahl, Anderson, and Libarkin 2005).

- Horizon Research Inc. conducted a study of middle school students' ideas related to plate tectonics. Several misconceptions about earthquakes and volcanoes were noted in their study, including (1) earthquakes caused Pangaea to break apart; (2) volcanoes and earthquakes always occur near plate boundaries, and they cannot occur in the interior of plates; (3) earthquakes are caused by plates crashing into each other, and the bigger the crash, the bigger the earthquake; (4) mountains form when earthquakes push the ground up; (5) earthquakes occur more frequently along coastlines with or without plate boundaries; (6) earthquakes cause all volcanic eruptions; and (7) volcanic eruptions cause all earthquakes. They also found that students recognized that earthquakes, volcanoes, and mountain formation usually occur in the same general areas, but students they

could not provide an explanation for it (Ford and Taylor 2006).

- Middle school students taught by traditional means are not able to construct coherent explanations about the causes of volcanoes and earthquakes (Duschl et al. 1992).

Suggestions for Instruction and Assessment

- Have students examine a world map of tectonic activity that shows the plates, earthquakes, and volcanoes. Have them look for evidence of volcanic and earthquake activity on each of the continents to support answer choice J. Use the same map to look for evidence that supports or refutes answer choices A, B, E, F, and I. The U.S. Geological Survey (USGS) produces an excellent map for this purpose, which can be downloaded from *http://pubs.usgs. gov/imap/2800*. The Smithsonian's Dynamic Planet website also contains an interactive map available at *http://nhb-arcims.si.edu/ ThisDynamicPlanet/index.html*.

- If students believe that earthquakes usually cause volcanic eruptions or vice versa, have students examine maps of earthquakes and volcanoes to see that although they follow a similar pattern in their location, some earthquakes occur without being near a volcano and vice versa. Also, students can access real-time earthquake data and compare earthquake sites with volcanic activity. The USGS has data available at *http://earthquake.usgs.gov/earthquakes* and *http://volcanoes.usgs.gov*.

- Follow the strand "Earthquakes and Volcanoes" on the Science Literacy Map for plate tectonics to examine the progression of ideas and connections. This map can be accessed at *http://strandmaps.dls.ucar. edu/?id=SMS-MAP-0049*.

- Browse the Digital Library for Earth System Education for additional resources on

earthquakes and volcanoes at *www.dlese. org/library/index.jsp.*

- Consider adding additional answer choices for high school students that would require them to account for hot spots, eruptions at mid-ocean ridges, or flood basalts.

References

Allen, M. 2010. *Misconceptions in primary science.* Berkshire, U.K.: Open University Press.

American Association for the Advancement of Science (AAAS). 2009. Benchmarks for science literacy online. *www.project2061.org/publications/bsl/online.*

Dahl, J., S. Anderson, and J. Libarkin. 2005. Digging into Earth science: Alternative conceptions held by K–12 teachers. *Journal of Science Education* 6 (2): 65–68.

Driver, R., A. Squires, P. Rushworth, and V. Wood-Robinson. 1994. *Making sense of secondary science: Research into children's ideas.* New York: RoutledgeFalmer.

Duschl, R., M. Smith, S. Kesidou, D. Gitomer, and L. Schauble. 1992. Assessing student explanations for criteria to format conceptual change learning environments. Paper presented at the annual meeting of the American Educational Research Association, San Francisco, CA.

Ford, B., and M. Taylor. 2006. Investigating students' ideas about plate tectonics. *Science Scope* 30 (1): 38–43.

National Research Council (NRC). 2012. *A framework for K–12 science education: Practices, crosscutting concepts, and core ideas.* Washington, DC: National Academies Press.

Section 4
Natural Resources, Pollution, and Human Impact

Concept Matrix..142

Related *Next Generation Science*
** *Standards* Performance Expectations** 143

Related NSTA Resources 143

28   **Renewable or Nonrenewable?**.................145
29   **Acid Rain** 151
30   **What Is a Watershed?**.............................155
31   **Is Natural Better?** 161
32   **The Greenhouse Effect** 165

Concept Matrix
Probes #28–#32

	#28 Renewable or Nonrenewable?	#29 Acid Rain	#30 What Is a Watershed?	#31 Is Natural Better?	#32 The Greenhouse Effect
PROBES					
GRADE RANGE →	**3–12**	**6–12**	**5–12**	**5–12**	**6–12**
CORE CONCEPTS ↓					
acid rain		X			X
agriculture				X	
energy resources	X				
global warming					X
greenhouse effect					X
natural resources	X			X	
nonrenewable resources	X				
ozone depletion					X
pollution		X		X	
renewable resources	X				
sustainability	X				
watershed			X		

Related *Next Generation Science Standards* Performance Expectations (NGSS Lead States 2013)

. . . .

Earth and Human Activity

- Grade 4, 4-ESS3-1: Obtain and combine information to describe that energy and fuels are derived from natural resources and their uses affect the environment.
- Grade 4, 4-ESS3-2: Generate and compare multiple solutions to reduce the impacts of natural Earth processes on humans.
- Grade 5, 5-ESS3-1: Obtain and combine information about ways individual communities use science ideas to protect the Earth's resources and environment.
- Grades 6–8, MS-ESS3-3: Apply scientific principles to design a method for monitoring and minimizing a human impact on the environment.
- Grades 6–8, MS-ESS3-4 : Construct an argument supported by evidence for how increases in human population and per-capita consumption of natural resources impact Earth's systems.
- Grades 6–8, MS-ESS3-5: Ask questions to clarify evidence of the factors that have caused the rise in global temperatures over the past century.
- Grades 9–12, HS-ESS3-1: Construct an explanation based on evidence for how the availability of natural resources, occurrence of natural hazards, and changes in climate have influenced human activity.
- Grades 9–12, HS-ESS3-3: Create a computational simulation to illustrate the relationships among management of natural resources, the sustainability of human populations, and biodiversity.
- Grades 9–12, HS-ESS3-4: Evaluate or refine a technological solution that reduces impacts of human activities on natural systems.
- Grades 9–12, HS-ESS3-6: Use a computational representation to illustrate the relationships among Earth systems and how those relationships are being modified due to human activity.

Reference

NGSS Lead States. 2013. *Next Generation Science Standards: For states, by states.* Washington, DC: National Academies Press. *www.nextgenscience.org/next-generation-science-standards.*

Related NSTA Resources
NSTA Press Books

Blake, R., J. Frederick, S. Haines, and S. Lee. 2010. *Inside-out: Environmental science in the classroom and the field, grades 3–8.* Arlington, VA: NSTA Press.

Carlson, W., N. Trautman, and The Environmental Inquiry Team. 2004. *Watershed dynamics.* Teacher ed. Arlington, VA: NSTA Press.

Environmental Literacy Council, and National Science Teachers Association. 2007. *Global climate change: Resources for environmental literacy.* Arlington, VA: NSTA Press.

Kastens, K., and M. Turrin. 2010. *Earth science puzzles: Making meaning from data.* Arlington, VA: NSTA Press.

Konicek-Moran, R. 2013. *Everyday Earth and space science mysteries: Stories for inquiry-based science teaching.* Arlington, VA: NSTA Press.

Soukhome, J., G. Peaslee, C. Van Faasen, and W. Statema. 2009. *Watershed investigations: 12 labs for high school science.* Arlington, VA: NSTA Press.

Trautmann, N., and The Environmental Literacy Team. 2003. *Decay and renewal.* Teacher ed. Arlington, VA: NSTA Press.

NSTA Journal Articles

Beckrich, A. 2014. The green room: Pests and pesticides. *The Science Teacher* 81 (4): 12.

Bednarski, M., and F. Holt. 2007. Science sample: The DEP saves the beans—A performance assessment task for acid rain. *Science Scope* 30 (8): 50–53.

Benson, R. 2003. Island watershed activity: Introducing students to watersheds and water quality. *The Science Teacher* 70 (2): 26–29.

Bodson, A., and L. Shive. 2004. Watershed investigations. *Science Scope* 27 (7): 21–23.

Damonte, K. 2004. Science shorts: Understanding acid rain. *Science and Children* 42 (3): 53.

Endreny, A. 2007. Watershed season*s*. *Science and Children* 44 (9): 20–25.

Greitz-Miller, R. 2006. Issues in depth: Inside global warming. *Science Scope* 30 (2): 56–60.

Likens, G. 2004. Science 101: What is acid rain? *Science and Children* 42 (3): 52.

McCallie, E. 2003. Science 101: What is a watershed? *Science and Children* 40 (7): 17.

Miller, R. 2005. Issues in depth: Pesticides, people, and the environment—A complex relationship. *Science Scope* 29 (2): 64–68.

Royce, C., and M. Holzer. 2003. What would it be like? Examining nonrenewable resources through inquiry-based activities. *The Science Teacher* 70 (4): 20–24.

Snyder, R. 2012. Science shorts: Truffula tree troubles. *Science and Children* 49 (8): 70–72.

Staehling, E. 2014. Clearing the air: Using probeware and online simulations to understand the greenhouse effect. *The Science Teacher* 82 (9): 50–56.

West, D., and D. Sterling. 2001. Soil studies: Applying acid-base chemistry to environmental analysis. *The Science Teacher* 68 (8): 37–40.

NSTA Learning Center Resources
NSTA Webinars

Get Energized: Interactive Generate! Game Explores Energy Choices and Environmental Quality
https://learningcenter.nsta.org/products/ symposia_seminars/EPA/webseminar2.aspx
List of NOAA Climate Change Webinars
http://learningcenter.nsta.org/products/web_ seminar_archive_sponsor.aspx?page=NOAA
NGSS Core Ideas: Earth and Human Activity
http://learningcenter.nsta.org/products/ symposia_seminars/NGSS/webseminar33.aspx

NSTA Science Objects

Resources and Human Impact: Environmental Degradation
http://learningcenter.nsta.org/ resource/?id=10.2505/7/SCB-RHI.3.1
Resources and Human Impact: Population Growth, Technology, and the Environment
http://learningcenter.nsta.org/ resource/?id=10.2505/7/SCB-RHI.2.1

Renewable or Nonrenewable?

People depend on Earth's resources for many things. Some of Earth's resources are renewable over a person's lifetime. Some are nonrenewable over a person's lifetime. Put an X next to all the things that are considered renewable resources over a person's lifetime.

_____ **A.** wind

_____ **B.** soil

_____ **C.** meat from cows

_____ **D.** water

_____ **E.** wood for building houses

_____ **F.** coal

_____ **G.** salmon

_____ **H.** granite rock for building material

_____ **I.** oxygen we breathe

_____ **J.** aluminum used in cans

_____ **K.** cotton to make clothing

_____ **L.** flowers

_____ **M.** naturally formed gemstones

_____ **N.** nuclear fuel

_____ **O.** corn

_____ **P.** petroleum oil

_____ **Q.** gold

_____ **R.** natural gas

_____ **S.** sunshine

_____ **T.** cooking oil

Explain your thinking. Describe what you know about renewable and nonrenewable resources.

_____

_____

_____

_____

_____

Renewable or Nonrenewable?

Teacher Notes

Purpose

The purpose of this assessment probe is to elicit students' ideas about Earth's natural resources. The probe is designed to find out if students can distinguish between renewable and nonrenewable natural resources.

Type of Probe

Justified list

Related Concepts

Energy resources, natural resources, nonrenewable resources, renewable resources, sustainability

Explanation

The best answer is A, C, D, E, G, I, K, L, O, S, and T. Earth's land, ocean, atmosphere, and biosphere provide humans with many things they need. Natural resources include air, water, soil, rocks and minerals, metals, energy, plants, and animals. These natural resources are considered renewable if they are replenished naturally and over relatively short periods of

time. Some renew at faster rates than others, making them more sustainable than those that do not renew very quickly. For instance, sunshine, which is used for solar energy, is renewable because the Sun always shines. Wind energy is another renewable resource. You cannot stop the wind from blowing any more than you can stop the Sun from shining, which makes wind easy to "renew."

Any plants that are grown for use in food and manufactured products are also renewable resources. Trees used for building material, cotton used for clothes, and food crops (such as corn, which can also be used to make a fuel called ethanol), can all be replanted and regrown after the harvest is collected. Animals are also considered a renewable resource because, like plants, animals can be bred to make more. Livestock (such as cows, pigs, and chickens) all fall into the renewable category. Fish are also considered renewable, but this categorization is a bit trickier because although some fish are actually farmed for production,

much of what we eat comes from wild stocks in lakes and the ocean. Those wild populations are in a delicate balance, and if that balance is upset by too much fishing, the population may die out. If a population dies out, it is no longer considered a renewable resource.

Depending on how water is used, it may be considered a renewable resource. You cannot really "use up" water, but you also cannot make more of it. There is a limited supply of fresh water on Earth, and it cycles through the planet in various forms—as a liquid (bodies of water, tiny droplets in clouds), a solid (polar ice caps and glaciers, tiny ice crystals in clouds), and a gas (water vapor). Liquid water can be used to generate hydroelectric power, which we get from water flowing through dams. Hydroelectric power is considered a renewable resource because we do not actually take the water out of the system and use it up to get electricity.

Some students may think oxygen is a nonrenewable resource because we use it up when we breathe. However, oxygen is continually cycled back into the atmosphere through photosynthesis.

Nonrenewable resources are not easily replenished by the environment. They can also be available in limited supplies. This limitation is usually due to the long time it takes for nonrenewable resources to be replenished. Replenishing an inch of soil may take 200 years or more. Some resources take thousands and millions of years to replenish. Nonrenewable resources include soil, ores containing metals, rocks and minerals, and energy fuels. The aluminum used to make beverage cans comes from an ore. The aluminum can be recycled from the can, but it cannot be renewed (replenished).

Administering the Probe

This probe can be used with grades 3–12. It can be combined with the card sort formative assessment classroom technique described on pages 4–5. Be aware that some students

may confuse this probe with renewable and nonrenewable energy resources in general. Make sure students know this probe is about Earth's natural resources, including living and nonliving materials and energy fuels. Make sure they know the time frame for this probe is whether a resource can be replenished in one's lifetime (estimate 100 years).

Related Core Ideas in *Benchmarks for Science Literacy* (AAAS 2009)

6–8 The Earth
- Some material resources are very rare and some exist in great quantities. The ability to obtain and process resources depends on where they are located and the form they are in. As resources are depleted, they may become more difficult to obtain.
- The wasteful or unnecessary use of natural resources can limit their availability for other purposes. Restoring depleted soil, forests, or fishing grounds can be difficult and costly.
- The benefits of Earth's resources—such as fresh water, air, soil, and trees—can be reduced by deliberately or inadvertently polluting them. The atmosphere, the oceans, and the land have a limited capacity to absorb and recycle waste materials.

9–12 The Earth
- The Earth has many natural resources of great importance to human life. Some are readily renewable, some are renewable only at great cost, and some are not renewable at all.

Related Core Ideas in *A Framework for K–12 Science Education* (NRC 2012)

3–5 ESS3.A: Natural Resources
- Energy and fuels that humans use are derived from natural sources, and their

use affects the environment in multiple ways. Some resources are renewable over time, and others are not.

6–8 ESS3.A: Natural Resources

- Humans depend on Earth's land, ocean, atmosphere, and biosphere for many different resources. Minerals, fresh water, and biosphere resources are limited, and many are not renewable or replaceable over human lifetimes. These resources are distributed unevenly around the planet as a result of past geologic processes.

Related Research

- Rule (2005) found that the use of the common word *oil* to refer to petroleum contributes to misconceptions. Some students think *oil* refers to cooking oil and thus may confuse renewable cooking oil with nonrenewable petroleum oil.
- In a study involving Turkish high school students, students held ideas such as (1) natural gas is a renewable resource, (2) renewable resources do not damage the environment, and (3) fossil fuels take only hundreds of years to replenish (Tortop 2012).

Suggestions for Instruction and Assessment

- This probe can be used as a card sort strategy (Keeley, Eberle, and Tugel 2007). Place the answer choices on cards and have students work in small groups to sort them into renewable resources and nonrenewable resources. Encourage students to discuss and defend their reasons for their card placement.
- Challenge students with more nuanced resources that may be both renewable *and* nonrenewable. An example would be lumber from trees. Trees can be planted, and therefore many people believe that they are completely renewable. This, however,

is often not the case when we think about the time frame for renewing a resource. If trees are clear-cut and far more trees are removed than can be grown in a lifetime, the resource is not renewed within a usable amount of time. Also, trees that take a very long time to grow to maturity, such as the giant redwoods, may not be renewed in one's lifetime.

- Water, trees, fish, crops, and other naturally occurring resources are vital for human survival. Many of these resources could be classified as nonrenewable if they are not used responsibly. For example, fishing in an area at a rate above that of the fishes' reproductive cycle would cause the resource to become scarce and possibly extinct, and thus not renewable. Combine this probe with uncovering ideas about sustainability and responsible use of natural resources.
- It takes 100 years or more to make an inch of soil, on average. Students may think soil is a renewable resource because it is so plentiful. You may ask your students to research the amount of time it takes for soil to form and report on situations in which soil depletion is a problem.
- Compare and contrast the terms *recyclable* and *renewable*. Tin and aluminum cans are made from nonrenewable resources but can be recycled back into tin and aluminum cans or into other products. Bottled water contains water, a renewable resource, but the container is made from a nonrenewable resource that is recyclable. Discuss the advantages and disadvantages of recycling nonrenewable resources.
- Challenge students to come up with other materials that are difficult to classify as simply renewable or nonrenewable.
- Have students create a story or comic book about a nonrenewable resource with a sequential description of the conditions

that existed to create the raw materials, the composition of the resource, the process that created the resource, the way it is extracted, the environmental damage caused by extraction (if any), the way it is used, and the amount of the resource that is left.

- Have students debate whether the 3 Rs—reduce, reuse, and recycle—should be expanded to the 4 Rs—reduce, reuse, recycle, and renew.

- The Environmental Protection Agency has a resource for teachers to help define and describe natural resources, renewable resources, and nonrenewable resources. The following website also provides fact sheets on resource usage: *www3.epa.gov/ epawaste/education/quest/pdfs/unit1/chap1/ u1_natresources.pdf.*

References

American Association for the Advancement of Science (AAAS). 2009. Benchmarks for science literacy online. *www.project2061.org/ publications/bsl/online.*

Keeley, P., F. Eberle, and J. Tugel. 2007. *Uncovering student ideas in science, vol. 2: 25 more formative assessment probes.* Arlington, VA: NSTA Press.

National Research Council (NRC). 2012. *A framework for K–12 science education: Practices, crosscutting concepts, and core ideas.* Washington, DC: National Academies Press.

Rule, A. 2005. Elementary students' ideas concerning fossil fuel energy. *Journal of Geoscience Education* 53 (3): 309–318.

Tortop, A. 2012. Awareness and misconceptions of high school students about renewable energy resources and applications: Turkey case. *Energy Education Science and Technology Part B: Social and Educational Studies* 4 (3): 1829–1840.

Acid Rain

Seven friends were talking about acid rain. They each had different ideas about acid rain. This is what they said:

Lynda: I've heard that we have acid rain because of the hole in the ozone layer.

Sam: All rain is acidic. This is normal.

Hank: The rain didn't used to be acidic. Global warming has made it acidic.

Otto: Pollutants evaporate from lakes, rivers, and ground spills and come down again as acid rain.

Elisa : Acid rain is caused when normal rain combines with things in the air that come from pollution.

Imani: Acid rain used to be a big problem. Due to regulations and laws, there is no more acid rain.

Paolo: Carbon monoxide from automobile exhaust combines with rain to make acid rain.

Which student do you agree with the most? _____ Explain your thinking.

_____

_____

_____

_____

Acid Rain

Teacher Notes

Purpose

The purpose of this assessment probe is to elicit students' ideas about pollution. The probe is designed to uncover commonly held ideas about acid rain.

Type of Probe

Friendly talk

Related Concepts

Acid rain, pollution

Explanation

The best answer is Elisa's: "Acid rain is caused when normal rain combines with things in the air that come from pollution." Oxides of sulfur and nitrogen are produced from industrial pollution, such as the burning of coal. These substances rise into the atmosphere, where they mix and react with water to form a more acidic pollutant, known as *acid rain*.

Normal rain is slightly acidic, with a pH of about 5.7. The rain that falls on land contains some dissolved carbon dioxide from the surrounding air. This causes rainwater to be slightly acidic because of the carbonic acid that is formed when pure water (pH 7) mixes with carbon dioxide. This process is beneficial in breaking down minerals found in the soil to make them accessible to plants for nutrient absorption. Acid rain resulting from reaction with pollutants has a pH of about 4. The lower the pH number, the more acidic the rain is.

Acid rain used to be a bigger problem than it is today. The worst effects in the United States were seen in the eastern states, which have more concentrated industrialization. In those areas, acid rain killed trees and harmed lakes and streams. Some lakes were declared biologically dead. The pH of rain in the northeast in the 1950s and 1960s was commonly as low as 4.0–4.5, with some storms delivering rain with a pH as low as 3.0. Compare that with the pH of vinegar, which is 2.4. Pollutants from the factory smokestacks in the Midwest are carried by the wind to the Northeast and

exacerbate the problem. In Europe, acid rain had a devastating effect on marble statues and buildings.

In the United States, the Clean Air Act of 1970 and the Acid Rain Program, established under Title IV of the 1990 Clean Air Act Amendments, aggressively worked to reduce emissions from power plants. The efforts resulted in the highly acidic rains in urban areas dropping to an average of pH 4.8 today. Unfortunately, we still experience harmful acidic rains and need more work to bring those pH levels down to a range that does not damage natural systems or the built environment, including marble statues and stone that reacts with acid.

Administering the Probe

This probe is best used with students in grades 6–12 who have some knowledge of the basic chemistry involved.

Related Core Ideas in *Benchmarks for Science Literacy* (AAAS 2009)

. . . .

6–8 The Earth

- The benefits of Earth's resources—such as fresh water, air, soil, and trees—can be reduced by deliberately or inadvertently polluting them. The atmosphere, the oceans, and the land have a limited capacity to absorb and recycle waste materials. In addition, some materials take a long time to degrade. Therefore, cleaning up polluted air, water, or soil can be difficult and costly.

9–12 The Earth

- Although the Earth has a great capacity to absorb and recycle materials naturally, ecosystems have only a finite capacity to withstand change without experiencing major ecological alterations that may also have adverse effects on human activities.

Related Core Ideas in *A Framework for K–12 Science Education* (NRC 2012)

. . . .

6–8 ESS3.C: Human Impacts on Earth Systems

- Human activities have significantly altered the biosphere, sometimes damaging or destroying natural habitats and causing the extinction of other species. But changes to Earth's environments can have different impacts (negative and positive) for different living things.

9–12 ESS3.C: Human Impacts on Earth Systems

- Scientists and engineers can make major contributions by developing technologies that produce less pollution and waste and that preclude ecosystem degradation.

Related Research

- Khalid (2001) found that elementary preservice teachers thought that pollutants evaporate and come down as acid rain. He also found a common misconception that the students thought acids have a higher pH than bases.
- Darcin (2010) found that 76% of the students sampled erroneously held the idea that carbon monoxide is responsible for acid rain. More than half (55%) saw carbon dioxide as a factor of more acid rain. Nearly half of the students (48%) thought that chlorofluorocarbons cause acid rain.
- Researchers have found that students of all ages are aware of environmentally "friendly" and "unfriendly" actions, and they know about a range of environmental problems. However, they tend not to link causes with their consequences and may have a tendency to imagine that all environmentally friendly actions help to solve all environmental problems (Driver et al. 1994).

Suggestions for Instruction and Assessment

- High school students can investigate the effects of a solution of simulated acid rain (4 ml of 1 M sulfuric acid in 2 L of distilled water) on various materials that might come in contact with acidic rain, such as brick, marble, concrete, assorted plant leaves, wood, different types of soils, metals, and plastics. Students can brainstorm what materials to test and bring in their own samples. The results can lead to a discussion about the possible effects of acid rain on the natural and built environment. (Safety notes: Have students wear sanitized goggles, nitrile gloves, and aprons throughout the activity. Have eyewash and shower stations within 10 seconds access in case of a splash. Review and share safety instructions for handling sulfuric acid. Be sure students wash their hands with soap and water after completing the activity.)

- Students can research the efforts being made to reduce the harmful effects of acid rain. A starting place for teachers is the following Environmental Protection Agency website: *www.epa.gov/acidrain/reducing.*

- Students can create their own experiments to test the effects of acid rain on plant germination. Obtain a variety of fast-germinating seeds to sprout in various environments: pure water, normal rain (captured on a rainy day), simulated solution of acid rain (4 ml of 1 M sulfuric acid in 2 L of distilled water), and diluted vinegar. (Safety notes: Have students wear sanitized goggles, nitrile gloves, and aprons throughout the activity. Have eyewash and shower stations within 10 seconds access in case of a splash. Review and share safety instructions for handling sulfuric acid. Be sure students wash their hands with soap and water after completing the activity.)

- Make connections to wind patterns to show how factory and coal plant emissions from one region in the United States (e.g., the Midwest) can affect the air and water in another region (e.g., New England).

References

American Association for the Advancement of Science (AAAS). 2009. Benchmarks for science literacy online. *www.project2061.org/publications/bsl/online.*

Darcin, S. 2010. Trainee science teachers' ideas about environmental problems caused by vehicle emissions. *Asia-Pacific Forum on Science Learning and Teaching* 11 (2): Article 14.

Driver, R., A. Squires, P. Rushworth, and V. Wood-Robinson. 1994. *Making sense of secondary science: Research into children's ideas.* London: RoutledgeFalmer.

Khalid, T. 2001. Preservice teachers' misconceptions regarding three environmental issues. *Canadian Journal of Environmental Education* 6 (1): 101–120.

National Research Council (NRC). 2012. *A framework for K–12 science education: Practices, crosscutting concepts, and core ideas.* Washington, DC: National Academies Press.

What Is a Watershed?

Five friends were talking about watersheds. They each had different ideas about what a watershed is. This is what they said:

Haru: I think a watershed is an undeveloped area of branching creeks, brooks, and streams that flow down a mountain or mountain range.

Penny: I think a watershed is a group of buildings and human-made structures that water drains off of and then collects in a certain area.

Ariana: I think a watershed is a tower or some type of building that stores water for human use.

Beau: I think a watershed is an area of land where all of the water that is under it or drains off of it goes into the same place.

Coco: I think a watershed is an underground area of soil or sand that holds water when it drains into the ground.

Which friend do you agree with the most? _____ Explain your thinking.

_____

_____

_____

_____

What Is a Watershed?

Teacher Notes

Purpose

The purpose of this assessment probe is to elicit students' ideas about watersheds. The probe is designed to find out what students think a watershed is.

Type of Probe

Friendly talk

Related Concepts

Watershed

Explanation

The best answer is Beau's: "I think a watershed is an area of land where all of the water that is under it or drains off of it goes into the same place." This answer choice is the same definition the Environmental Protection Agency (EPA) uses to define a watershed. Furthermore, scientist and geographer John Wesley Powell described a watershed as "that area of land, a bounded hydrologic system, within which all

living things are inextricably linked by their common water course and where, as humans settled, simple logic demanded that they become part of a community" (EPA 2013).

Watersheds come in all shapes and sizes. They include both developed and undeveloped land areas. They cross county and state boundaries. In the continental United States, there are 2,110 watersheds.

Understanding the watershed concept is important for managing water resources and understanding issues about water quality, point and nonpoint source pollution, land use and rapid urban development, groundwater contamination, and the impact of personal actions and behavior on water quality.

Administering the Probe

This probe is best used with grades 5–12. The probe can be extended by using the annotated drawing formative assessment classroom technique, which is described on pages 4–5.

Related Core Ideas in *Benchmarks for Science Literacy* (AAAS 2009)

6–8 The Earth

- Fresh water, limited in supply, is essential for some organisms and industrial processes. Water in rivers, lakes, and underground can be depleted or polluted, making it unavailable or unsuitable for life.

Related Core Ideas in *A Framework for K–12 Science Education* (NRC 2012)

3–5 ESS3.C: Human Impacts on Earth Systems

- Human activities in agriculture, industry, and everyday life have had major effects on the land, vegetation, streams, ocean, air, and even outer space. But individuals and communities are doing things to help protect Earth's resources and environments.

6–8 ESS2.C: The Roles of Water in Earth's Surface Processes

- Water continually cycles among land, ocean, and atmosphere via transpiration, evaporation, condensation and crystallization, and precipitation, as well as downhill flows on land.

6–8 ESS3.C: Human Impacts on Earth Systems

- Human activities have significantly altered the biosphere, sometimes damaging or destroying natural habitats and causing the extinction of other species. But changes to Earth's environments can have different impacts (negative and positive) for different living things.

9–12 ESS3.C: Human Impacts on Earth Systems

- The sustainability of human societies and the biodiversity that supports them requires responsible management of natural resources.

Related Research

- To elicit students' ideas about watersheds, a study was conducted that asked students to draw a picture of a watershed and explain their drawing. Results of the study showed students understand a watershed from a very limited scientific perspective. Several sixth- and seventh-grade students described and drew a watershed as a water storage facility or a facility that supplies water. Eighth- and ninth-grade students in the study focused on a mountainous stream. Older students also incorporated the water cycle, but they rarely represented links between land and watercourses. For all students in the study, humans did not appear to be a part of a watershed; they were regarded as separate from it (Shepardson, Harbor, and Wee 2005).

- A study by Patterson and Harbor (2005) investigated the effect of a watershed curriculum on students' geoscience learning. They found that some students thought of watersheds as water towers or well houses. They concurred that because of our everyday language, containing the word *shed* seems to imply some kind of structure with a roof that holds water.

- Studies on watersheds suggest that children associate a watershed with a mountainous, rustic area, rather than a more developed region (Shepardson et al. 2007).

- Middle school students' ideas about watersheds are not much different than those of adults. Studies suggest that science education is contributing little to the development of a citizenship knowledgeable about watersheds. Most citizens are not knowledgeable about the watershed concept, nor do they fully understand the hydrologic connection (Schueler 2000).

- A sample study showed that only 41% of the adults surveyed had any idea about what a watershed is and only 22% knew

that storm water runoff was a major cause of stream pollution (NEETF 1998).

Suggestions for Instruction and Assessment

- Have students make annotated drawings of a watershed. Drawings give students who have difficulty constructing verbal or written explanations an opportunity to reveal and explain their ideas through visual means (Keeley 2008). Having students explain their drawing reveals additional ideas they have about watersheds.

- Findings from Shepardson, Harbor, and Wee's study (2005, pp. 384–385) suggest that the following concepts need to be developed to improve students' understandings of watersheds and to enhance students' geoscience learning:

  1. A watershed is a land area that provides runoff that feeds rivers and streams.

  2. Every place on land is a part of a watershed, including the places where we live, work, and play.

  3. Smaller streams flow into larger rivers forming a river system, which is a network of tributaries that flow into a major river.

  4. Watersheds consist of a river system, which drains water from the land within the watershed.

  5. Watershed boundaries are defined based on elevation. The elevation or divide determines the direction water flows or into which basin precipitation flows.

  6. Earth's surface consists of numerous nested and joining watersheds that drain into lakes or the ocean.

  7. Sediment and other substances and contaminants on land are transported into the stream through runoff and then transported by the river system through the watershed into joining watersheds, lakes, or oceans. The contaminants

transported off the land area and through the river system are often referred to as *nonpoint source pollution*. Fertilizer and pesticide runoff, for example, are nonpoint source pollutants.

- Use topographic maps or maps with elevations to have students discover that not all watersheds include mountains or mountain ranges with high elevations.

- Students can gain an intuitive understanding of the physical aspects of watersheds by creating their own watershed models out of a crumpled piece of paper. Conclude by having them use a map to trace the watershed that supplies their drinking water. More activity information is available at *www.omsi.edu/sites/all/FTP/files/expeditionnw/4.E.1.Crumple.pdf*. This activity supports the scientific practice of developing and using models to explain ideas. Be sure to discuss aspects of the model that represent a real watershed and those that do not represent a real watershed.

- Take students on a field trip to a local watershed and identify land use practices that affect the water quality in the watershed area.

References

American Association for the Advancement of Science (AAAS). 2009. *Benchmarks for science literacy online*. *www.project2061.org/publications/bsl/online*.

Keeley, P. 2008. *Science formative assessment: 75 practical strategies for linking assessment, instruction, and learning*. Thousand Oaks, CA: Corwin Press.

National Environmental Education Training Foundation (NEETF). 1998. National report card on environmental knowledge, attitudes, and behaviors: Seventh annual Roper survey of adult Americans. NEETF, Washington, DC.

National Research Council (NRC). 2012. *A framework for K–12 science education: Practices,*

crosscutting concepts, and core ideas. Washington, DC: National Academies Press.

Patterson, L., and J. Harbor. 2005. Using assessment to evaluate and improve inquiry-based geo-environmental science activities: Case study of a middle school watershed *E. coli* investigation. *Journal of Geoscience Education* 53 (2): 204–214.

Schueler, T. 2000. On watershed education. In *The Practice of Watershed Protection*, ed. T. Schueler and K. Holland, 629–635. Ellicott City, MD: Center for Watershed Protection.

Shepardson, D., J. Harbor, and B. Wee. 2005. Water towers, pump houses, and mountain streams: Students' ideas about watersheds. *Journal of Geoscience Education* 53 (4): 381–386.

Shepardson, D., B. Wee, M. Priddy, L. Schellenberger, and J. Harbor. 2007. What is a watershed? Implications of student conceptions of environmental science education and the National Science Education Standards. *Science Education* 91 (4): 554–578.

U.S. Environmental Protection Agency (EPA). 2013. Water resources. EPA. *http://water.epa.gov/type/watersheds/whatis.cfm.*

Is Natural Better?

Two friends were talking about how people use pesticides and fertilizers to grow their vegetables. They each had different ideas. This is what they said:

Joaquim: We should use natural fertilizers and pesticides since they are not harmful to the environment like chemical fertilizers and pesticides.

Kaylan: I disagree with you. Natural fertilizers and pesticides can also harm the environment.

Who do you agree with the most? _____ Explain your thinking.

_____

_____

_____

_____

_____

_____

_____

_____

Is Natural Better?

Teacher Notes

Purpose

The purpose of this assessment probe is to elicit students' ideas about natural versus human-engineered products. The probe is designed to uncover whether students think natural always means not harmful.

Type of Probe

Opposing views

Related Concepts

Agriculture, natural resources, pollution

Explanation

The best answer is Kaylan's: "I disagree with you. Natural fertilizers and pesticides can also harm the environment." All chemicals, natural and commercially manufactured, can be harmful if used in high enough quantities—even water, which we consider essential to life. "The dose makes the poison" is a rule that applies to all chemical compounds, natural or manufactured. It is a common misconception that if something is "natural," it is safe for humans and the environment.

Students may think that natural pesticides are not harmful. Some of them can be quite toxic, not only to insects but also to other organisms that you do not want to harm, such as pets and wildlife. Rotenone, nicotine, and pyrethrum are examples of natural, botanical insecticides. Rotenone is produced from members of the bean plant family. It has been used as a crop insecticide since the mid-1800s to control leaf-eating caterpillars. It is six times more toxic than carbaryl, which is a synthetic product used for the same purpose. Nicotine sulfate has been used since the turn of the century and is the most hazardous botanical insecticide available to home gardeners. The insecticide is extracted from tobacco and is highly toxic to humans and other warm-blooded animals. It is six times more toxic than diazinon, which is a widely available synthetic insecticide sold for controlling many of the same pests.

Pyrethrum, extracted from the dried flowers of the pyrethrum daisy, has a rapid "knockdown" effect on many insects. It has very low toxicity to mammals, but it breaks down quickly, thus limiting the amount of time that it is effective. One of the benefits of natural pesticides is that they break down rapidly in the environment and often pose no additional threat in runoff that makes its way to creeks, rivers, and the ocean.

Natural fertilizers can be harmful to the environment, especially water resources. Excess nutrients from fertilizers can run into our lakes, ponds, rivers, streams, and even the ocean, causing algae blooms. When the algae die, they sink to the bottom and decompose in a process called *eutrophication*, which removes oxygen from the water. Fish and other aquatic species cannot survive in these dead zones.

All technologies, including chemically engineered products, have benefits and risks. Whether a pesticide or fertilizer is natural or engineered by humans, they all contain chemicals that should be used responsibly.

Administering the Probe

This probe is best used with grades 5–12. This probe can be combined with the lines of agreement formative assessment classroom technique, which is discussed in the classroom snapshot described on pages 6–9.

Related Core Ideas in *Benchmarks for Science Literacy* (AAAS 2009)

3–5 Agriculture

- Damage to crops by rodents, weeds, or insects can be reduced by using poisons, but their use may harm other plants or animals.

6–8 Agriculture

- In agriculture, as in all technologies, there are always trade-offs to be made.

9–12 Agriculture

- Agricultural technology requires trade-offs between increased production and environmental harm and between efficient production and social values.

Related Core Ideas in *A Framework for K–12 Science Education* (NRC 2012)

3–5 ESS3.C: Human Impacts on Earth Systems

- Human activities in agriculture, industry, and everyday life have had major effects on the land, vegetation, streams, ocean, air, and even outer space. But individuals and communities are doing things to help protect Earth's resources and environments.

6–8 ESS3.C: Human Impacts on Earth Systems

- Human activities have significantly altered the biosphere, sometimes damaging or destroying natural habitats and causing the extinction of other species. But changes to Earth's environments can have different impacts (negative and positive) for different living things.

- Typically as human populations and per-capita consumption of natural resources increase, so do the negative impacts on Earth unless the activities and technologies involved are engineered otherwise.

9–12 ESS3.C: Human Impacts on Earth Systems

- Scientists and engineers can make major contributions by developing technologies that produce less pollution and waste and that preclude ecosystem degradation.

Related Research

- Researchers have found that students of all ages are aware of environmentally "friendly" and "unfriendly" actions, and they know about a range of environmental problems.

However, they tend not to link causes with their consequences and may have a tendency to imagine that all environmentally friendly actions help solve all environmental problems (Driver et al. 1994).

Suggestions for Instruction and Assessment

- Students can conduct research to gather information on the safety and effectiveness of commercial versus natural pesticides. Hold a discussion or debate during which students can make claims and defend them with evidence.
- Have students design a fair test to determine whether commercial fertilizer or natural fertilizer is more beneficial for their selected plants. (Safety notes: Have students wear sanitized goggles, nitrile gloves, and aprons throughout the activity. Have eyewash and shower stations within 10 seconds access in case of a splash. Review and share safety instructions for handling sulfuric acid. Be sure students wash their hands with soap and water after completing the activity.)
- Students can design a system for capturing the runoff from commercial fertilizer and natural fertilizer. Have students apply their systems to selected plants to determine if the runoff can be harmful to plants.
- Students can work in groups to create a continuum of natural materials that range from not harmful to somewhat harmful to

very harmful. Have each group describe their reasoning for placing each item at its specific location on the continuum.
- The "Ban DHMO" internet hoax was first started by students at the University of California, Santa Cruz, to convince people that dihydrogen monoxide (the chemical name for water) was a dangerous chemical that should be banned. The information they shared is true but easily misinterpreted from a harmful viewpoint. Share the website *http://dhmo.org* with students to show how easily it is to be persuaded that anything with a "chemical name" is harmful and how easy it is to misinterpret information, even when the information is true.
- Contact your local Master Gardeners program for in-depth information on natural versus chemical fertilizers and pesticides. View a list of those programs at *www.ahs. org/gardening-resources/master-gardeners*.

References

American Association for the Advancement of Science (AAAS). 2009. Benchmarks for science literacy online. *www.project2061.org/publications/bsl/online*.

Driver, R., A. Squires, P. Rushworth, and V. Wood-Robinson. 1994. *Making sense of secondary science: Research into children's ideas*. London: RoutledgeFalmer.

National Research Council (NRC). 2012. *A framework for K–12 science education: Practices, crosscutting concepts, and core ideas*. Washington, DC: National Academies Press.

The Greenhouse Effect

Many people have heard of "the greenhouse effect." But what does that mean? Put an X next to the statements you think apply to the greenhouse effect.

_____ **A.** The greenhouse effect is related to increasing global temperatures.

_____ **B.** The greenhouse effect supports why we should stop building greenhouses.

_____ **C.** The greenhouse effect is about the thinning of the ozone layer.

_____ **D.** The greenhouse effect contributes to increased incidences of skin cancer.

_____ **E.** The greenhouse effect reduces the amount of oxygen in the atmosphere.

_____ **F.** The greenhouse effect is caused by using spray cans and air conditioners.

_____ **G.** The greenhouse effect is the same thing as global warming.

_____ **H.** The greenhouse effect is the main cause of hurricanes.

_____ **I.** The greenhouse effect can contribute to a change in weather patterns.

_____ **J.** The greenhouse effect can be reduced by using unleaded gasoline.

_____ **K.** The greenhouse effect is related to increased use of fossil fuels.

_____ **L.** The greenhouse effect is one of the causes of acid rain.

_____ **M.** The greenhouse effect is related to human activities.

_____ **N.** The greenhouse effect can be controlled by keeping beaches clean.

Explain your thinking. Describe what you know about the greenhouse effect.

_____

_____

_____

The Greenhouse Effect

Teacher Notes

Purpose

The purpose of this assessment probe is to elicit students' ideas about the greenhouse effect. The probe is designed to reveal whether students confuse the greenhouse effect with other environmental concerns or lump the concerns together.

Type of Probe

Justified list

Related Concepts

Acid rain, global warming, greenhouse effect, ozone depletion

Explanation

The best answer is A, I, K, and M. Life on Earth has evolved with varying amounts of greenhouse gases in the atmosphere. A greenhouse gas is capable of absorbing infrared radiation, thereby trapping and holding heat in the atmosphere. By increasing the heat in the atmosphere, greenhouse gases are responsible

for the greenhouse effect. The most common greenhouse gases are carbon dioxide (CO_2) and water vapor. Other common greenhouse gases are methane, nitrous oxide, ozone, and fluorocarbons. Greenhouse gases have the ability to reflect infrared light back to Earth, thus warming the planet. If there were no greenhouse gases in our atmosphere, Earth would be very cold. You can imagine the greenhouse gases in our atmosphere acting like the windows in a car. They let visible light through, but hold the reflected infrared light in the car, which warms the car up. Greenhouses function the same way, but students most likely will have more experience being in a car that warms up, even on a cold day, if the windows are closed.

At the onset of the Industrial Age (1880s), fossil fuels were mined from below the surface of Earth to use for heat, manufacturing, and transportation. When burned, they released additional CO_2 into the atmosphere, causing Earth to become warmer on average. As more CO_2 is added to the atmosphere, more heat is

reflected back to Earth, like adding blankets on a bed. This trend has continued, with the 10 hottest years since 1880 occurring in the past 17 years. Concentrations of atmospheric CO_2 have increased from an average of 278 parts per million (ppm) over the past 800,000 years, to 300 ppm in 1950, to 403 ppm in 2015. Human activity is the largest source of CO_2 emissions. Burning fossil fuel to provide energy for heat, light, manufacturing, and transportation contributes the most CO_2 to the atmosphere. Additional contributors are agriculture and deforestation. Natural sources of CO_2 include volcanic eruptions and decaying plant matter, but these are in small amounts as compared with human-caused sources. Human activity is related to the greenhouse effect we experience today, but students must also know that the greenhouse effect existed long before human activity. The issue today is the significantly increased amounts of greenhouse gases that result from human activity.

As Earth warms, the increased energy is dissipated in numerous ways, resulting in hot days becoming hotter, rainfall and flooding becoming heavier, hurricanes becoming stronger, and droughts becoming more severe.

The greenhouse effect is often confused with global warming, although the two are related. The *greenhouse effect* is the name scientists have given to Earth's ability to retain heat. When the Sun's rays reach the planet, approximately two-thirds of the thermal energy enters Earth's atmosphere and is absorbed by the planet's surface. Earth then emits this thermal energy, which is absorbed by the atmosphere. The atmosphere radiates the heat back toward Earth, which absorbs the heat again. This process keeps the planet warm and can almost be thought of like a blanket covering Earth. *Global warming* is an increase in Earth's overall temperature. As more greenhouse gases accumulate in the atmosphere, Earth warms because more radiated heat is trapped.

Administering the Probe

This probe is best used with grades 6–12. This probe can be combined with the card sort formative assessment classroom technique, which is described in the classroom snapshot described on pages 6–9.

Related Core Ideas in *Benchmarks for Science Literacy* (AAAS 2009)

9–12 The Earth

- Greenhouse gases in the atmosphere, such as carbon dioxide and water vapor, are transparent to much of the incoming sunlight but not to the infrared light from the warmed surface of the Earth. When greenhouse gases increase, more thermal energy is trapped in the atmosphere, and the temperature of the Earth increases the light energy radiated into space until it again equals the light energy absorbed from the Sun.

Related Core Ideas in *A Framework for K–12 Science Education* (NRC 2012)

6–8 ESS3.C: Human Impacts on Earth Systems

- Typically as human populations and per-capita consumption of natural resources increase, so do the negative impacts on Earth unless the activities and technologies involved are engineered otherwise.

6–8 ESS3.D: Global Climate Change

- Human activities, such as the release of greenhouse gases from burning fossil fuels, are major factors in the current rise in Earth's mean surface temperature (global warming). Reducing the level of climate change and reducing human vulnerability to whatever climate changes do occur depend on the understanding of climate

science, engineering capabilities, and other kinds of knowledge, such as understanding of human behavior and on applying that knowledge wisely in decisions and activities.

9–12 ESS2.D: Weather and Climate

- Changes in the atmosphere due to human activity have increased carbon dioxide concentrations and thus affect climate.

- Current models predict that, although future regional climate changes will be complex and varied, average global temperatures will continue to rise. The outcomes predicted by global climate models strongly depend on the amounts of human-generated greenhouse gases added to the atmosphere each year and by the ways in which these gases are absorbed by the ocean and biosphere.

Related Research

- Boyes and Stanisstreet (1993) found that some middle school students have acceptable ideas about global warming, including the notion that an increase in greenhouse gases will change weather patterns.

- Research shows that some students tend to lump environmental issues together. Driver et al. (1994) lists several misconceptions among students ages 11–16 that contribute to their confusing greenhouse effect and global warming with other environmental problems, including (1) the use of lead-free gas reduces global warming, (2) ozone layer depletion is the same as the greenhouse effect, and (3) the greenhouse effect can be reduced by keeping beaches clean and protecting wildlife.

- In one study, the greenhouse gases conceptions of fifth-grade students from diverse languages and cultures revealed two patterns. First, students erroneously equated Earth's greenhouse effect with the literal definition of a greenhouse. Because of this misunderstanding, they often described inadequate amounts of oxygen

and the need to stop building greenhouses. Second, students erroneously equated the greenhouse effect with the ozone layer, identifying aerosols and air conditioners as causing holes in the ozone layer, which, in turn, cause more cases of skin cancer (Lee et al. 2007).

- Rye, Rubba, and Wiesenmayer (1997) studied the conceptions of global warming and greenhouse gases held by 24 American students in grades 6–8, following their completion of a global warming unit. Their analysis of students' concept maps showed that about 50% of the students maintained the alternative conceptions that ozone layer depletion is a key cause of global warming and that CO_2 destroys the ozone layer.

Suggestions for Instruction and Assessment

- Have students design an experiment that simulates the greenhouse effect. Various systems could be explored. One example is a car on a sunny day. Students can take temperature readings of the car in the shade for a baseline reading. Then, they can move the car to a sunny spot and, with the windows rolled up, take incremental temperature readings. Another example models the greenhouse effect using glass jars—one with no cover and one with plastic film sealing the jar. Students can put thermometers in the jars, place the jars in a sunny spot, and take incremental temperature readings.

- Students can research the CO_2 content in the atmosphere of Venus and compare its extreme greenhouse effect with that of Earth.

- Computer models and simulations can contribute to your students' understanding of the greenhouse effect. The National Academies of Sciences has a video available

at *http://goo.gl/Ck65bh*. The University of Colorado, Boulder, has an interactive simulation of the greenhouse effect available at *http://phet.colorado.edu/en/simulation/greenhouse*.

- Demonstrate the greenhouse effect with an interactive game. This model incorporates both constructive and destructive human activities that effect the amount of greenhouse gases in our atmosphere: *www.pbs.org/strangedays/educators/season1/ag_odf_cogame.html*.

- Be aware that some students think global warming is the same as the greenhouse effect. This may be because the two are often mentioned together and, as a result, students use them interchangeably.

References

American Association for the Advancement of Science (AAAS). 2009. Benchmarks for science literacy online. *www.project2061.org/publications/bsl/online*.

Boyes, E., and M. Stanisstreet. 1993. The "greenhouse effect": Children's perceptions of causes, consequences and cures. *International Journal of Science Education* 15 (5): 531–552.

Driver, R., A. Squires, P. Rushworth, and V. Wood-Robinson. 1994. *Making sense of secondary science: Research into children's ideas*. London: RoutledgeFalmer.

Lee, O., B. Lester, O. Ma, J. Lambert, and M. Jean-Baptiste. 2007. Conceptions of the greenhouse effect and global warming among elementary students from diverse languages and cultures. *Journal of Geoscience Education* 55 (2): 117–125.

National Research Council (NRC). 2012. *A framework for K–12 science education: Practices, crosscutting concepts, and core ideas*. Washington, DC: National Academies Press.

Rye, J., P. Rubba, and R. Wiesenmayer. 1997. An investigation of middle school students' alternative conceptions of global warming. *International Journal of Science Education* 19 (5): 527–551.

Index

A

A-B-C-C-B-V strategy, 56

A Framework for K–12 Science Education, xv

  Acid Rain probe and, 153

  Are They Talking About Climate or Weather? probe and, 75

  Can a Plant Break Rocks? probe and, 105

  Coldest Winter Ever! probe and, 71

  Describing Earth's Plates probe and, 127

  Does the Ocean Influence Our Weather or Climate? probe and, 67

  Grand Canyon probe and, 109

  The Greenhouse Effect probe and, 167–168

  Groundwater probe and, 35

  How Do Rivers Form? probe and, 119

  How Many Oceans and Seas? probe and, 39

  How Old Is Earth? probe and, 89

  Is It a Fossil? probe and, 93

  Is It Erosion? probe and, 101

  Is Natural Better? probe and, 163

  Land or Water? probe and, 27

  Mountains and Beaches probe and, 113

  Renewable or Nonrenewable? probe and, 147–148

  Sedimentary Rock Layers probe and, 97

  Water Cycle Diagram probe and, 51

  Weather Predictors probe and, 59

  What Are the Signs of Global Warming? probe, 79–80

  What Do You Know About Soil? and, 21

  What Do You Know About Volcanoes and Earthquakes? probe and, 137–138

  What Is a Watershed? probe and, 157

  What Is the Inside of Earth Like? probe and, 123

  What's Beneath Us? probe and, 17

  Where Did the Water in the Puddle Go? probe and, 55

  Where Do You Find Earth's Plates? probe and, 133

  Where Is Most of the Fresh Water? probe and, 31

  In Which Direction Will the Water Swirl? probe and, 63

  Why Is the Ocean Salty? probe and, 43

A Private Universe video, ix

Acid Rain probe, 151–154

  administration of, 153

  explanation of, 152–153

  purpose of, 152

  related concepts for, 142, 152

  related core ideas in science education standards, 153

  research related to, 153

  suggestions for instruction and assessment using, 154

  Teacher Notes for, 152–154

  type of, 152

Acid Rain Program, 153

Administering probes, xiv

American Association for the Advancement of Science (AAAS), 105, 128, 129. *See also Benchmarks for Science Literacy*

American Museum of Natural History, 97

Annotated drawing strategy, 4–5, 16, 50, 123, 156, 158

Are They Talking About Climate or Weather? probe, 73–76

  administration of, 74

  explanation of, 74

  purpose of, 74

  related concepts for, 46, 74

  related core ideas in science education standards, 75

  research related to, 75

  suggestions for instruction and assessment using, 75–76

  Teacher Notes for, 74–76

  type of, 74

Argumentation, xviii, 5, 7–9

Atlas of Science Literacy, xv

Ault, C., 97

Index

B

Barr, V., 55

Benchmarks for Science Literacy, xv

  Acid Rain probe and, 153

  Are They Talking About Climate or Weather? probe and, 75

  Can a Plant Break Rocks? probe and, 105

  Coldest Winter Ever! probe and, 71

  Describing Earth's Plates probe and, 127

  Does the Ocean Influence Our Weather or Climate? probe and, 67

  Grand Canyon probe and, 109

  The Greenhouse Effect probe and, 167

  Groundwater probe and, 34

  How Do Rivers Form? probe and, 119

  How Many Oceans and Seas? probe and, 39

  How Old Is Earth? probe and, 89

  Is It a Fossil? probe and, 93

  Is It Erosion? probe and, 101

  Is Natural Better? probe and, 163

  Land or Water? probe and, 26

  Mountains and Beaches probe and, 113

  Renewable or Nonrenewable? probe and, 147

  Sedimentary Rock Layers probe and, 96–97

  Water Cycle Diagram probe and, 51

  Weather Predictors probe and, 59

  What Are the Signs of Global Warming? probe, 79

  What Do You Know About Soil? and, 21

  What Do You Know About Volcanoes and Earthquakes? probe and, 137

  What Is a Watershed? probe and, 157

  What Is the Inside of Earth Like? probe and, 123

  What's Beneath Us? probe and, 17

  Where Did the Water in the Puddle Go? probe and, 55, 56

  Where Do You Find Earth's Plates? probe and, 133

  Where Is Most of the Fresh Water? probe and, 30–31

  In Which Direction Will the Water Swirl? probe and, 63

  Why Is the Ocean Salty? probe and, 43

Boyes, E., 168

Broadwell, B., 89

C

Can a Plant Break Rocks? probe, 103–106

  administration of, 104

  explanation of, 104

  purpose of, 104

  related concepts for, 84, 104

  related core ideas in science education standards, 105

  research related to, 105

  suggestions for instruction and assessment using, 105–106

  Teacher Notes for, 104–106

  type of, 103

Card sort strategy, 3–4, 21, 22, 74, 75, 80, 93, 100, 101, 127, 137, 147, 148, 167

Clean Air Act (1970) and Clean Air Act Amendments (1990), 153

Climate. *See* Global warming; Weather and climate

Climate Generation, 80

Climate Literacy: The Essential Principles of Climate Science, 71, 76

Coldest Winter Ever! probe, 69–72

  administration of, 70

  explanation of, 70

  purpose of, 70

  related concepts for, 46, 70

  related core ideas in science education standards, 71

  research related to, 71

  scenario of use in high school classroom, 6–9

  suggestions for instruction and assessment using, 71–72

  Teacher Notes for, 70–72

  type of, 70

Concept matrix, xiii, xv

  for Earth History, Weathering and Erosion, and Plate Tectonics probes, 84

  for Land and Water probes, 12

  for Natural Resources, Pollution, and Human Impact probes, 142

  for Water Cycle, Weather, and Climate probes, 46

Index

Conceptual change, xvi, xviii, 2, 5
Coriolis effect, 62, 63

D
Darcin, S., 153
Describing Earth's Plates probe, 125–129
    administration of, 127
    explanation of, 126–127
    purpose of, 126
    related concepts for, 84, 126
    related core ideas in science education
        standards, 127
    research related to, 127–128
    scenario of use in middle school
        classroom, 5–6
    suggestions for instruction and
        assessment using, 128–129
    Teacher Notes for, 126–129
    type of, 126
Diagnostic assessment, xv, 2
Digital Library for Earth System Education,
    138–139
Disciplinary core ideas, xv, 1, 4
Distracters, xi, xv, 1
Does the Ocean Influence Our Weather or
    Climate? probe, 65–68
    administration of, 67
    explanation of, 66–67
    purpose of, 66
    related concepts for, 46, 66
    related core ideas in science education
        standards, 67
    research related to, 67
    suggestions for instruction and
        assessment using, 68
    Teacher Notes for, 66–68
    type of, 66
Driver, R., xv, 168

E
Earth History, Weathering and Erosion, and
    Plate Tectonics probes, 83–139
    Can a Plant Break Rocks?, 103–106
    concept matrix for, 84
    Describing Earth's Plates, 125–129
    Grand Canyon, 107–110
    How Do Rivers Form?, 117–120
    How Old Is Earth?, 87–90

Is It a Fossil?, 91–94
Is It Erosion?, 99–102
Mountains and Beaches, 111–115
NSTA resources related to, 85–86
related *NGSS* performance expectations
    for, 85
Sedimentary Rock Layers, 95–98
What Do You Know About Volcanoes and
    Earthquakes?, 135–139
What Is the Inside of Earth Like?, 121–
    124
Where Do You Find Earth's Plates?,
    131–134
El Niño weather pattern, 59, 68
Environmental Protection Agency (EPA),
    149, 154, 156
Erosion, xvii, 17, 21, 42, 85, 89. *See also*
    Weathering
    Grand Canyon probe, 108, 109
    How Do Rivers Form? probe, 118–120
    Is It Erosion? probe, 99–102
    Mountains and Beaches probe, 112–114
Evaporation, xii, xvii, 34, 43, 44, 67, 119,
    151, 153, 157
    Water Cycle Diagram probe, 50–51
    Where Did the Water in the Puddle Go?,
        54–56
Evolution, 89, 93

F
Familiar phenomenon probes, 2, 54
The Farmer's Almanac, 57, 58, 60
Feller, R., 43, 67
Fertilizers, 158
    Is Natural Better? probe, 161–164
Formative assessment classroom
    techniques (FACTs), xiv, 2, 3, 7
Formative assessment probes, xi–xviii
    administration of, xiv
    definition of, 1
    distracters for, xi, xv, 1
    Earth and environmental science–related
        in other books in *Uncovering Student
        Ideas in Science* series, xi–xiii, xvi
        scenarios of use of, 3–9
        in *Science and Children,* xvi–xvii
    explanation of, xiv
    features of, 1–3

Index

format of, xiii–xvi
grade levels for use of, 1, 2
language used in, xi, 1
professional development for use of, 2–3,
   9
purpose of, xi, xiv, xviii
references for, xvi
related concepts for, xiv
related core ideas for, xv
research related to, xv
resources about, xiii, xiv, xi–xiii, xviii, 9
suggestions for instruction and
   assessment using, xv–xvi
Teacher Notes for, xiii–xvi, 1
types of, xiv, 1–2
use in the elementary classroom, xvi–xvii
Freyberg, P., 43, 113
Friendly talk probes, xiv, 2, 4, 16, 26, 30, 34,
   38, 42, 50, 54, 58, 62, 66, 88, 104, 108,
   112, 118, 122, 132, 152, 156

G
Geology
   Can a Plant Break Rocks? probe, 103–
      106
   Describing Earth's Plates probe, 125–129
   Grand Canyon probe, 107–110
   How Do Rivers Form? probe, 117–120
   How Old Is Earth? probe, 87–90
   Is It a Fossil? probe, 91–94
   Mountains and Beaches probe, 111–115
   Sedimentary Rock Layers probe, 95–98
   What Do You Know About Soil? probe,
      19–23
   What Do You Know About Volcanoes and
      Earthquakes? probe, 135–139
   What Is the Inside of Earth Like? probe,
      121–124
   What's Beneath Us? probe, 15–18
   Where Do You Find Earth's Plates?
      probe, 131–134
Global warming, xii, 8, 73, 151
   Coldest Winter Ever! probe, 70, 72
   The Greenhouse Effect probe, 165–169
   What Are the Signs of Global Warming?
      probe, 77–80
Gosselin, D., 22
Grand Canyon probe, 107–110

administration of, 109
explanation of, 108
purpose of, 108
related concepts for, 84, 108
related core ideas in science education
   standards, 109
research related to, 109–110
suggestions for instruction and
   assessment using, 110
Teacher Notes for, 108–110
type of, 108
Greenhouse effect
   The Greenhouse Effect probe, 165–169
   What Are the Signs of Global Warming?
      probe, 78–80
The Greenhouse Effect probe, 165–169
   administration of, 167
   explanation of, 166–167
   purpose of, 166
   related concepts for, 142, 166
   related core ideas in science education
      standards, 167–168
   research related to, 168
   suggestions for instruction and
      assessment using, 168–169
   Teacher Notes for, 166–169
   type of, 166
Groundwater probe, 33–35
   administration of, 34
   explanation of, 34
   purpose of, 4, 34
   related concepts for, 12, 34
   related core ideas in science education
      standards, 34–35
   research related to, 35
   scenario of use in fifth-grade classroom,
      4–5
   suggestions for instruction and
      assessment using, 35
   Teacher Notes for, 34–35
   type of, 34

H
Happs, J., 17, 97, 114
Harbor, J., 157, 158
Hattie, John, 1
Horizon Research Inc. (HRI), 128, 133, 138
How Do Rivers Form? probe, 117–120

Index

administration of, 118–119
explanation for, 118
purpose of, 118
related concepts for, 84, 118
related core ideas in science education
   standards, 119
research related to, 119
suggestions for instruction and
   assessment using, 119–120
Teacher Notes for, 118–120
type of, 118
How Many Oceans and Seas? probe, 37–40
administration of, 39
explanation of, 38–39
purpose of, 38
related concepts for, 12, 38
related core ideas in science education
   standards, 39
research related to, 39
suggestions for instruction and
   assessment using, 39
Teacher Notes for, 38–40
type of, 38
How Old Is Earth? probe, 87–90
administration of, 88–89
explanation of, 88
purpose of, 88
related concepts for, 84, 88
related core ideas in science education
   standards, 89
research related to, 89
suggestions for instruction and
   assessment using, 89–90
Teacher Notes for, 88–90
type of, 88

I

In Which Direction Will the Water Swirl?
   probe, 61–63
administration of, 62
explanation of, 62
purpose of, 62
related concepts for, 46, 62
related core ideas in science education
   standards, 63
research related to, 63
suggestions for instruction and
   assessment using, 63

Teacher Notes for, 62–63
type of, 62
Intergovernmental Panel on Climate
   Change, 74
Is It a Fossil? probe, 91–94
administration of, 92–93
explanation of, 92
purpose of, 92
related concepts for, 84, 92
related core ideas in science education
   standards, 93
research related to, 93
suggestions for instruction and
   assessment using, 93
Teacher Notes for, 92–94
type of, 92
Is It Erosion? probe, 99–102
administration of, 100
explanation of, 100
purpose of, 100
related concepts for, 84, 100
related core ideas in science education
   standards, 101
research related to, 101
suggestions for instruction and
   assessment using, 101–102
Teacher Notes for, 100–102
type of, 100
Is Natural Better? probe, 161–164
administration of, 163
explanation of, 162–163
purpose of, 162
related concepts for, 142, 162
related core ideas in science education
   standards, 163
research related to, 163–164
suggestions for instruction and
   assessment using, 164
Teacher Notes for, 162–164
type of, 162

J
Jones, M., 89
Justified list probes, xiv, 2, 3, 20, 74, 78, 92,
   96, 100, 126, 136, 146, 166

K
Keeley, P., ix, xvi, xxiii–xxiv

Index

Khalid, T., 153

L

La Niña weather pattern, 59, 68
Land and Water probes, 11–44
    concept matrix for, 12
    Groundwater, 33–35
    How Many Oceans and Seas?, 37–40
    Land or Water?, 25–28
    NSTA resources related to, 13–14
    related *NGSS* performance expectations
      for, 13
    What Do You Know About Soil?, 19–23
    What's Beneath Us?, 15–18
    Where Is Most of the Fresh Water?,
      29–31
    Why Is the Ocean Salty?, 41–44
Land or Water? probe, 25–28
    administration of, 26
    explanation of, 26
    purpose of, 26
    related concepts for, 12, 26
    related core ideas in science education
      standards, 26–27
    research related to, 27
    suggestions for instruction and
      assessment using, 27
    Teacher Notes for, 26–28
    type of, 26
Learning goals, xv
Learning process, xvi, 1
Lines-of-agreement strategy, 7
Literacy connections, 2
Lyell, C., 89

M

Macklem-Hurst, J., 22
Magnuson, J., 75
Making Sense of Secondary Science:
    Research Into Students' Ideas, xv
Metacognition, xviii, 2
Misunderstandings and misconceptions, ix,
    xi, xvi, 1, 2, 9. *See also specific probes*
Models, development and use of, xviii, 2.
    See also specific probes
Mountains and Beaches probe, 111–115
    administration of, 113
    explanation of, 112

    purpose of, 112
    related concepts for, 84, 112
    related core ideas in science education
      standards, 113
    research related to, 113–114
    suggestions for instruction and
      assessment using, 114
    Teacher Notes for, 112–115
    type of, 112

N

NASA, 31, 44, 75–76
National Academies of Sciences, 168
National Science Foundation, 60
National Science Teachers Association
    (NSTA) resources, xiii, xiv, xviii, 9. *See*
    also references for specific probes
    for Earth History, Weathering and
      Erosion, and Plate Tectonics probes,
      85–86
    for Land and Water probes, 13–14
    for Natural Resources, Pollution, and
      Human Impact probes, 143–144
    NSTA Learning Center, xvi
    NSTA Safety Portal, xvi
    Science and Children, xv, xvi–xvii
    Science Scope, 5
    for Water Cycle, Weather, and Climate
      probes, 47–48
Natural Resources, Pollution, and Human
    Impact probes, 141–169
    Acid Rain, 151–154
    concept matrix for, 142
    The Greenhouse Effect, 165–169
    Is Natural Better?, 161–164
    NSTA resources related to, 143–144
    related *NGSS* performance expectations
      for, 143
    Renewable or Nonrenewable?, 145–149
    What Is a Watershed?, 155–159
Natural Resources of Canada, 124
Next Generation Science Standards
    (NGSS), xv, 2, 4
    performance expectations, xiii, xv, 4
      for Earth History, Weathering and
        Erosion, and Plate Tectonics
        probes, 85, 119
      for Land and Water probes, 13

for Natural Resources, Pollution, and Human Impact probes, 143
for Water Cycle, Weather, and Climate probes, 47
NOAA, 44, 68, 75–76, 80

O

Ocean Literacy Framework, 39, 68
Opposing views probes, xiv, 7, 70, 162
Ozone layer depletion, 151, 165, 168

P

Patterson, L., 157
PBS Learning Media, 75, 110, 128
Performance expectations. *See Next Generation Science Standards*
Pesticides, 22, 98, 158
    Is Natural Better? probe, 161–164
Phenomena and Representations for the Instruction of Science in Middle Schools (PRISMS), 128–129
Plate tectonics, xi, 123
    Describing Earth's Plates probe, 5–6, 125–129
    Science Literacy Map for, 138
    What Do You Know About Volcanoes and Earthquakes? probe, 135–138
    Where Do You Find Earth's Plates? probe, 131–134
Pollution, 147, 156, 158, 162, 163
    Acid Rain probe, 151–154
Powell, J. W., 156
Predict-explain-observe (P-E-O) technique, 27
Principles of Geology, 89
Professional development, 2–3, 9
Project 2061, 105, 128

R

Recycling, 147, 148, 149, 153
Renewable or Nonrenewable? probe, 145–149
    administration of, 147
    explanation of, 146–147
    purpose of, 146
    related concepts for, 142, 146
    related core ideas in science education standards, 147–148
    research related to, 148
    suggestions for instruction and assessment using, 148–149
    Teacher Notes for, 146–149
    type of, 146
Research related to probes, xv
River Cutters curriculum, 120
Rubba, P., 168
Rule, A., 148
Rye, J., 168

S

Safety notes, xvi. *See also specific probes*
Science and Children, xv, xvi–xvii
Science Formative Assessment: 50 More Practical Strategies for Linking Assessment, Instruction, and Learning, xviii, 9
Science Formative Assessment: 75 Practical Strategies for Linking Assessment, Instruction, and Learning, xviii, 9
Science Scope, 5
Scientific literacy, xviii
Scripps Institution of Oceanography, 133
Seas of the World, 39
Sedimentary Rock Layers probe, 95–98
    administration of, 96
    explanation of, 96
    purpose of, 96
    related concepts for, 84, 96
    related core ideas in science education standards, 96–97
    research related to, 97
    suggestions for instruction and assessment using, 97–98
    Teacher Notes for, 96–98
    type of, 96
Shepardson, D., 158
Smithsonian Institute, 22, 138
Soil Science Society of America, 22
Stanisstreet, M., 168
Stream tables, 101, 114, 119–120
Summative assessment, xii, xv
Sustainability, 143, 146, 148, 157

T

Taylor, A., 89
Teacher Notes, xiii–xvi, 1

Index

Teaching for Conceptual Understanding in Science, xviii, 9

This Dynamic Earth: The Story of Plate Tectonics, 128

Travis, A., 55

Tucker, L., xxiv–xxv

U

Uncovering Student Ideas in Science series, ix, xi, xviii, 1, 9

  Earth and environmental science–related probes in other books in, xi–xiii

University Corporation for Atmospheric Research, 71

U.S. Geological Survey (USGS), 31, 124, 128, 133, 138

W

Water

  Does the Ocean Influence Our Weather or Climate? probe, 65–68

  Groundwater probe, 4–5, 33–35

  How Do Rivers Form? probe, 117–120

  How Many Oceans and Seas? probe, 37–40

  Land or Water? probe, 25–28

  Water Cycle Diagram probe, 49–52

  What Is a Watershed? probe, 155–159

  Where Did the Water in the Puddle Go? probe, 53–56

  Where Is Most of the Fresh Water? probe, 29–31

  In Which Direction Will the Water Swirl? probe, 61–63

  Why Is the Ocean Salty? probe, 41–44

Water Cycle, Weather, and Climate probes, 45–81

  Are They Talking About Climate or Weather?, 73–76

  Coldest Winter Ever!, 69–72

  concept matrix for, 46

  Does the Ocean Influence Our Weather or Climate?, 65–68

  NSTA resources related to, 47–48

  related *NGSS* performance expectations for, 47

  Water Cycle Diagram, 49–52

  Weather Predictors, 57–60

What Are the Signs of Global Warming?, 77–81

Where Did the Water in the Puddle Go?, 53–56

In Which Direction Will the Water Swirl?, 61–63

Water Cycle Diagram probe, 49–52

  administration of, 50–51

  explanation of, 50

  purpose of, 50

  related concepts for, 46, 50

  related core ideas in science education standards, 51

  research related to, 51

  suggestions for instruction and assessment using, 51

  Teacher Notes for, 50–52

  type of, 50

Weather and climate

  Are They Talking About Climate or Weather? probe, 73–76

  Coldest Winter Ever! probe, 6–9, 69–72

  Does the Ocean Influence Our Weather or Climate? probe, 65–68

  Weather Predictors probe, 57–60

  What Are the Signs of Global Warming? probe, 77–81

Weather Predictors probe, 57–60

  administration of, 59

  explanation of, 58–59

  purpose of, 58

  related concepts for, 46, 58

  related core ideas in science education standards, 59

  research related to, 59

  suggestions for instruction and assessment using, 59–60

  Teacher Notes for, 58–60

  type of, 58

Weather Underground, 72

Weathering, xii, xvii, 17, 42, 85. *See also* Erosion

  Can a Plant Break Rocks? probe, 104, 105

  Grand Canyon probe, 108, 109

  How Do Rivers Form? probe, 119

  Is It Erosion? probe, 100–102

  Mountains and Beaches probe, 112, 113

Index

What Do You Know About Soil? probe, 20, 21

Wee, B., 158

What Are the Signs of Global Warming? probe, 77–81
  administration of, 79
  explanation of, 78–79
  purpose of, 78
  related concepts for, 46, 78
  related core ideas in science education standards, 79–80
  research related to, 80
  suggestions for instruction and assessment u sing, 80
  Teacher Notes for, 78–81
  type of, 78

What Are They Thinking? Promoting Elementary Learning Through Formative Assessment, xv, xvii, 9

What Do You Know About Soil? probe, 19–23
  administration of, 21
  explanation of, 20–21
  purpose of, 20
  related concepts for, 12, 20
  related core ideas in science education standards, 21
  research related to, 21–22
  scenario of use in third-grade classroom, 3–4
  suggestions for instruction and assessment using, 22
  Teacher Notes for, 20–23
  type of, 20

What Do You Know About Volcanoes and Earthquakes? probe, 135–139
  administration of, 137
  explanation of, 136–137
  purpose of, 136
  related concepts for, 84, 136
  related core ideas in science education standards, 137–138
  research related to, 138
  suggestions for instruction and assessment using, 138–139
  Teacher Notes for, 136–139
  type of, 136

What Is a Watershed? probe, 155–159
  administration of, 156
  explanation of, 156
  purpose of, 156
  related concepts for, 142, 156
  related core ideas in science education standards, 157
  research related to, 157–158
  suggestions for instruction and assessment using, 158
  Teacher Notes for, 156–159
  type of, 156

What Is the Inside of Earth Like? probe, 121–124
  administration of, 122
  explanation of, 122
  purpose of, 122
  related concepts for, 84, 122
  related core ideas in science education standards, 123
  research related to, 123
  suggestions for instruction and assessment using, 123–124
  Teacher Notes for, 122–124
  type of, 122

What's Beneath Us? probe, 15–18
  administration of, 16
  explanation of, 16
  purpose of, 16
  related concepts for, 12, 16
  related core ideas in science education standards, 17
  research related to, 17
  suggestions for instruction and assessment using, 17
  Teacher Notes for, 16–18
  type of, 16

Where Did the Water in the Puddle Go? probe, 53–56
  administration of, 54
  explanation of, 54
  purpose of, 54
  related concepts for, 46, 54
  related core ideas in science education standards, 55
  research related to, 55
  suggestions for instruction and assessment using, 55–56
  Teacher Notes for, 54–56

Index

type of, 54

Where Do You Find Earth's Plates? probe, 131–134
administration of, 132
explanation of, 132
purpose of, 132
related concepts for, 84, 132
related core ideas in science education standards, 133
research related to, 133
suggestions for instruction and assessment using, 133
Teacher Notes for, 132–134
type of, 132

Where Is Most of the Fresh Water? probe, 29–31
administration of, 30
explanation of, 30
purpose of, 30
related concepts for, 12, 30
related core ideas in science education standards, 30–31

research related to, 31
suggestions for instruction and assessment using, 31
Teacher Notes for, 30–31
type of, 30

Why Is the Ocean Salty? probe, 41–44
administration of, 43
explanation of, 42–43
purpose of, 42
related concepts for, 12, 42
related core ideas in science education standards, 43
research related to, 43
suggestions for instruction and assessment using, 43–44
Teacher Notes for, 42–44
type of, 42

Wiesenmayer, R., 168

Y

Yale Project on Climate Change Communication, 71, 75, 80